Study Guide for STATISTICS FOR MANAGEMENT

IBM/PC VERSION

SECOND EDITION

Study Guide for STATISTICS FOR MANAGEMENT

IBM/PC VERSION

SECOND EDITION

by LINCOLN CHAO
California State University, Long Beach

Prepared by

Grover Rodich
Portland State University

George A. Johnson
Idaho State University

John Kilpatrick
College of Idaho

for

P.S. Associates
Brookline, Massachusetts

The Scientific Press
540 University Avenue
Palo Alto, CA 94301-1985
(415) 322-5221

STUDY GUIDE FOR
STATISTICS FOR MANAGEMENT

Copyright © 1984 by The Scientific Press.
A revision of *Study Guide for Statistics for Management*
Copyright © 1980 by Wadsworth, Inc.

All rights reserved. No part of this book may be
reproduced, stored in a retreival system, or transcribed
in any form or by any means without the prior written
permission of the publisher, The Scientific Press.

Printed in the United States of America.

10 9 8 7 6 5 4 3 2 1

ISBN-0-89426-040-5

PREFACE

Not long after a student begins the study of the functional areas of business, such as accounting, finance, marketing, production, and personnel, the importance of data and information processing is realized. The term "data processing" has historically been associated with repetitive bookkeeping and clerical tasks and the use of the computer immediately comes to mind. Statistics, however, is also a disciplined study that can be classified as a means of processing data to gain insight into managerial problems that increase the effectiveness of decision making.

This student study guide is designed for the beginning business student to introduce him or her to the concept of statistical analysis and to assist in gaining an appreciation for statistical techniques as applied to business. The guide is designed for use with any basic statistics text such as *Statistics for Management*, Second Edition, by Lincoln Chao. Of particular interest to the student are the short computer programs written in BASIC at the end of key chapters. The programs were developed on an IBM Personal Computer and are available from the publisher on a floppy disk. (Write Scientific Press, 540 University Avenue, Palo Alto, CA 94301 or call 415 - 322-5221.) They were intentionally kept short and simple to make it easy to replicate them on any available computer system. No prior computer knowledge is necessary to utilize these programs. Students who have a fundamental knowledge of BASIC may want to modify the output section of the programs, and the authors feel that any study or modification of the programs would be valuable in instilling further insight into the statistical manipulations involved. If a floppy disk is used, be sure the first operation is to create a backup disk. This will prevent future sorrows.

Each chapter is structured with an identical format with most chapters containing

- A set of behavioral *Learning Objectives* for the student to use as a preview of the concepts to be learned. The objectives can also be used after finishing the chapter as a test of learning effectiveness.

- A *Synopsis* of the chapter with important concepts, procedures, and techniques presented and outlined following the order of their full description in the Chao text.

- A section listing the *Key Terms* presented in the chapter with a short definition for each term.

- A collection of important chapter formulas in a *Formula Review* section.

- A self-*Organized Learning Quiz* with *Answers* that is designed to lead the student from easier concepts to the more difficult. Open-ended questions and answers that the authors have found are most often asked by their own students in a classroom environment are given. By providing complete answers to these questions, the student has immediate feedback on his or her solution ability.

- A *Sample Exercise* is worked through to completion with comments and illustrations for each important step. An attempt is made to make solution procedures similar for several different types of problems so that the student can more easily understand concept similarities.

- A more involved *Case Problem* is presented to give the student a simulation of an actual application of the chapter tools. Some of the cases have solutions presented in the form of a computer output and others are intentionally left for the student to solve.

- For the chapters where applicable, a listing of a BASIC *Computer Program* is given with representative output. The authors have found these programs extremely useful for students of statistics. The programs are general in their design and can be used for any generated, empirical data input. A catalog of these programs on a computer system relieves much of the computational drudgery associated with a statistics class and allows the student to better appreciate the relationships, concepts, and, at times, the aesthetic qualities present in the field of applied statistics.

GROVER RODICH

GEORGE A. JOHNSON

JOHN KILPATRICK

CONTENTS

1	Statistics and Its Uses	1
2	Data Collection and Presentation	8
3	Data Description: Sample Statistics	19
4	Events and Probabilities	34
5	Probability Rules and Functions	46
6	Expected Value and Population Parameters	56
7	Discrete Probability Distributions	65
8	The Normal Probability Distribution	79
9	Statistical Estimation	93
10	Hypothesis Testing	107
11	Student's t Tests	118
12	Chi-Square Tests	134
13	Analysis of Variance	151
14	Elements of Modern Decision Theory	168
15	Simple Regression and Correlation	180
16	Multiple Regression and Correlation	196
17	Time Series	215
18	Nonparametric Methods	232
19	Statistical Quality Control	248

1
STATISTICS AND ITS USES

CHAPTER LEARNING OBJECTIVES

On completing this chapter, the student should be able to

- Differentiate between statistical statements derived from descriptive statistical methods and those derived from statistical inference.

- Describe the attributes of a population and a sample.

- Distinguish a population as being either finite or infinite in size.

- List the major fields in which statistical methods have played a crucial role.

- Explain areas in which a misuse of statistics can give deceptive and misleading information.

SYNOPSIS

The Study of Statistics

The student who begins the study of statistics often approaches it as another course in mathematics. This situation proabaly occurs because, as a discipline, statistics is concerned with quantitative data involving many mathematical symbols. It is not long after the initial contact, however, that the student finds that the study of statistics is unique and quite different from the ordinary course in mathematics.

Instead of being a deductive course similar to most quantitative courses, statistics is built on a logic process of induction and inference.

Basic Concepts

Decision makers, faced with an uncertain environment, strive to gain as much information relevant to the decision as possible. In this way, uncertainty is reduced and the problem solver feels more comfortable and confident with his or her eventual decision and actions. When a large mass of data must be analyzed as to its properties, an investigator often samples from the total and notes the properties of that sample. By inductive logic, the properties of the large population are then inferred. Properties of the sample are measured by using techniques studied in descriptive statistics.

A property partially describing the nature of a sample is called a *statistic*, whereas a property describing the nature of a population is called a *parameter*.

Populations can be classified as being either *infinite* or *finite*. An infinite population contains more units or observations than can be listed or counted. An unlimited amount is possible. Even though the number of grains of sand on all the world's beaches could theoretically be counted, practicality dictates it as being an infinite set. Also, if after observing an item it is returned and again available for observation, the number of observations becomes infinite even though the number of items observed is countable.

A population that contains a reasonably limited number of units is classified as a finite population.

Uses and Misuses of Statistics

Statistical techniques to process data into useful information for more effective decision making are employed by governmental agencies, research institutions, business and industry, and other organizations for strategic and developmental reasons. Data processed by statistical procedures become more valuable for assessing the environment of a firm, both internally and externally. Long-range planning is enhanced by statistical measures of the economic, political, social, and competitive setting of the organization. Specific uses include statistical demand forecasting, statistical quality control, inventory control, production control, market analysis, financial risk analysis, auditing procedures, bidding, and purchasing analysis, to name a few. Actually, it can play a vital role in almost all phases of business, government, and industry.

Care must be taken when using any kind of analysis in that assumptions should be clearly understood and procedural steps carefully followed. In using statistical methods, common pitfalls include improper definition of the population, biased sampling techniques, results that are misleading when graphed or presented in a deceptive manner, and reliance on results obtained with insufficient sample size.

KEY TERMS

Data — Groups of numerical symbols representing quantities, actions, or things. Data are the raw material for the statistical process.

Descriptive statistics	The discipline of collecting, classifying, and presenting numerical data.
Finite population	A population that is made of a limited number of measurements or observations
Inferential statistics	Techniques by which generalizations are made based on partial information gained by using descriptive techniques.
Infinite population	A population that is made of a large number of measurements or observations that cannot be entirely reached by counting.
Information	Data that have been processed into a form that is meaningful and valuable to the recipient.
Parameters	Measurable characteristics of a population, usually considered as true values.
Population	The totality of all potential observations or measurements under consideration in a given problem situation.
Sample	A collection of observations or measurements taken from a given population.
Statistic	A measurable characteristic of a sample. Sample statistics are used to make inferences about population parameters.
Statistics	An area of study containing methods of collecting, classifying, and presenting numerical data. Its purpose is also to make generalizations about the characteristics of all potential numerical data under consideration.

ORGANIZED LEARNING QUIZ

Of 1000 items in a warehouse, 300 are selected at random and 20 are found to be defective. A total inspection of the 1000 items produces 71 defective items.

1. The 1000 items constitute

    a. a sample statistic.
    b. the sample size.
    c. the population size.
    d. a population statistic.

2. The 300 items selected at random constitute

    a. the population size.
    b. the sample size.
    c. a sample parameter.
    d. a population parameter.

3. The 20 defective items in the group of 300 constitute

   (a.) a sample statistic.
   b. a population statistic.
   c. the population size.
   d. the sample size.

4. The 71 defective items constitute

   (a.) a population parameter.
   b. a sample parameter.
   c. the sample size.
   d. the population size.

5. The registrar of a small college has published a report that claims 43% of the student enrollment is female. A friend of yours attends this college and has told you that 75% of the students in her classes are female.

   (a.) Both percentages could be correct.
   b. One of the percentages must be wrong.
   c. Both percentages must be wrong.
   d. The percentages given must total 100%.

6. If the college's total enrollment is considered the population and the friend's classes a sample

   a. 43% is a statistic.
   (b.) 75% is a statistic.
   c. there are more boys than girls in the sample.
   d. there are more girls than boys in the population.

7. Making statements about a characteristic of a population based on observations of a similar characteristic in a sample is called _____ statistics.

8. A statistic is to a sample as a _____ is to a population.

9. A fruit grader pulls boxes of fruit off a wagon load in a random manner to sample the load. He will use _____ statistics to make inferences about the characteristics of the wagon load.

10. Gathering evidence and presenting it in a court of law represent an analogous example of the use of _____ statistics, whereas using this evidence to decide on guilt or innocence is an analogous example of _____ statistics.

11. Explain how a jar with five different-colored blocks could be classified as an infinite population.

    _____

    _____

    _____

12. Is the Dow Jones Industrial average considered a parameter

or a statistic? Explain.
_____
_____
_____

13. Is the Gross National Product considered a parameter or a statistic? Explain.
_____
_____
_____

14. An often-misused descriptive statistic is an average. Comment on the statement "The average family income in my neighborhood is $75,000 per year."
_____
_____
_____

15. What basic statistics error is a doctor making when he infers that everyone in the world must be sick after he spends a long day at his office?
_____
_____
_____

ANSWERS TO ORGANIZED LEARNING QUIZ

1. c   2. b
3. a   4. a
5. a   6. b
7. inferential   8. parameter
9. inferential   10. descriptive, inferential

11. Sampling one at a time with replacement in this manner could possibly never end; thus, one is sampling from an infinite population.

12. The Dow Jones Industrial Average is a descriptive measure of a representative sample of industrial firms; thus it is a statistic.

13. The GNP is a descriptive measure of a composite of all the economy; thus it is most likely to be used as a parameter.

14. If the average has been determined by adding the incomes of all families and then dividing by the number of families, one large income earner in a small neighborhood can greatly

destroy the total. As an example, consider $10,000, $10,000, $10,000, and $130,000. The average is $40,000--four times the majority of incomes.

15. He is making inferential statements based on obviously biased incomplete information.

SAMPLE EXERCISE

There are 750 workers at a large plant. It is desired to learn if the average distance (in miles) a worker travels from home to work is less than 5 miles. A sample of 50 workers is chosen by an investigator; and the average distance traveled by these 50 is 4.83 miles.

At this point the investigator makes the statement "The average distance that a worker in the plant drives to work is likely to be less than 5 miles."

The population is (finite, infinite).

4.83 miles is a (parameter, statistic).

It is the result of a (descriptive, inferential) statistics method.

The investigator's statement is based on (descriptive, inferential) statistics.

Assume that the true average value for all workers in the plant is really 5.7 miles. Comment.

_____

_____

_____

ANSWERS TO SAMPLE EXERCISE

In this exercise the population of interest is limited to the 750 workers at the plant. No attempt is to be made to make any inference about a larger set of workers. Therefore the population is of such a limited number that it can be classified as finite.

4.83 is a measurement characteristic of the sample; hence it is a statistic and the result of a descriptive statistics method.

The investigator's statement seems reasonable. The difference between the true value of the population parameter and the related measurement of the sample is called the sampling error. In this situation, the inferential statement was based on incomplete information and the sampling error caused a bad decision. Even under the most careful experimental conditions, such situations can occur. Statistics does not purport to prove hypothetical statements, only to lend support or nonsupport to their truth.

The large sampling error could have been caused by the investigator not controlling the selection procedure for his sample observation. Instead of a random sample, all the members of a particular department might have been chosen as the sample. The characteristics of those persons (age group, salary group, level of training) might have seriously biased the results. There is also the outside chance of picking a random sample and getting an abnormally large number of members in the sample who live close to the plant.

CASE PROBLEM

The Cue Family Department Store

The Cue Family department store is located in the county seat of the rural county of Westfir. The Cue family opened their first store in 1923 and presently occupy their fourth building. The business has suffered and prospered along with Westfir's economy through the years; it is now the largest and most complete store within a wide area.

The ladies' wear department has been a favorite among the women in the county, for the merchandise has always been complete, reasonably priced, and has matched the needs of the residents. Jane Cue is the present manager of that department; and since she was born and raised in Westfir County, she knows most of her customers.

Ms. Cue is particularly concerned about the pricing of items under her responsibility and is always quick to notice when demand falls off for reasons she attributes to a selling price that is too high.

One of the areas Ms. Cue pays particular attention to is ladies' suits. She always wants to know how the average price paid for a suit changes from week to week; in order to find out, she collects the data in Exhibit A.

The collection of data is difficult and Ms. Cue is not even sure whether it is accurate or complete. The data are actually only a good estimate and she wonders if there might be a better way to determine or estimate the average price.

Exhibit A
Prices Paid for All Women's Suits
Sold During Previous Week

76	93	103	92	95	90	98
86	102	86	97	93	89	93
92	88	91	83	104	92	88
83	93	87	90	89	94	97
94	88	95	79	105	90	92
103	86	91	87	95	92	97

2
DATA COLLECTION AND PRESENTATION

CHAPTER LEARNING OBJECTIVES

On completing this chapter, the student should be able to

- Distinguish a population as being either finite or infinite.

- Describe the procedures used to utilize simple random sampling, systematic sampling, stratified sampling, and cluster sampling.

- Describe the conditions that must exist when sampling for the process to be considered random sampling.

- Tabulate a set of numerical data into a frequency and cummulative frequency table.

- Exhibit qualitative data in various ways, including pie charts, pictographs, and bar graphs.

- Exhibit a set of numerical data in the form of histograms, frequency polygons, and ogive graphs.

SYNOPSIS

Random Sampling

Care must be taken by an investigator when drawing samples from a population for inferential purposes. Any set of n observations from a population is called a sample of size n; however, in order to have the sample contain as many of the characteristic attributes of the population, a *random* sample of size n is chosen.

Observations can be taken of populations that are either finite or infinite in number. For instance, sampling from a population and returning the sample to the population after each observation can theoretically continue forever. A sample from such an infinite population is considered a random sample if each observation in the sample does not affect the probability (chance) that any other observation will be selected. A sample from a finite population is considered random if every *available* potential observation in the population has the same probability of being selected. An observation is also referred to as a sampling unit.

Simple Random Sampling

If it is possible to attach a unique number on every possible sampling unit and, after thoroughly mixing them up, to choose one randomly, then the process is called *simple random sampling*. An alternate means of selection can use a table of random digits. By continuing down the table (or across), numbers are randomly generated and, in turn, will select the sampling units in a simple random sampling plan. As many digits in the table are taken as needed each time to ensure that every sampling unit has a chance of being selected. Care should be taken when using the table that a definite procedure is followed in choosing the numbers, such as top to bottom and side to side. If the procedure is changed or reversed, then randomness may not be ensured.

Systematic Sampling

If it is believed that samples close together tend to be homogeneous, whereas samples far apart may not be, then a systematic sample may be used by selecting every kth unit of the population. This step ensures the breakup of homogeneous partners. k is determined by dividing the population size by the sample size. The number of the first unit is selected by simple random sampling; then every kth one after that becomes a member of the sample.

Stratified Sampling

When homogeneous units are known to exist prior to sampling, they can be grouped into classes called *strata*. Simple random sampling techniques are then used on each stratum to generate small subsamples. These subsamples are next combined to form the overall desired sample.

A proportionate stratified sample is created when the number of units in each subsample is proportionate to the number of units in each stratum in the population. If one stratum contains 15% of the population, then its subsample will constitute 15% of the overall sample.

Disproportionate stratified sampling is based on the premise that a group of largely homogeneous units does not need a large sample to determine its characteristics. Thus the subsample sizes are inversely related to the homogeneity of the units in the stratum. This technique reduces the cost of sampling.

Cluster Sampling

Cluster sampling procedures require the random selection of small groups (clusters) of units as a first step. The overall sample is then made up either of all the units of the clusters or of a subsample of each cluster. The clusters are referred to as primary sampling units; when combined to form an overall sample, the procedure is single-stage sampling. Subsampling creates a two-stage sampling procedure; and if three or more stages are used, it is called multi-stage sampling. Cost reduction is the motivation behind cluster sampling procedures.

Data Organization

Once a random sample is drawn, a means of organizing and presenting the data is necessary to make any analysis easier to perform.

Array. Raw data put in either numerical ascending order or descending order (sequenced) are said to be in an array.

Frequency Distribution. By grouping data into classes, the number of occurrences of observations in each class will form a frequency distribution. Class intervals are usually formed according to the following guidelines: (1) the number of classes should not be fewer than six or greater than 20; (2) the range of the classes should be the same; (3) the range should be an odd number; and (4) the midpoints of the classes should have the same number of digits as the raw data. The last guideline will put the endpoints of each class between observed values, thereby ensuring that none of the observed values will lie on the endpoints.

Raw data are more manageable when put in a frequency distribution. The assumptions are that each class can be represented by its midpoint and the observations in each class are evenly distributed about the midpoint. Even though usually not true, the discrepancies of the classes tend to offset each other. The investigator must weigh the advantages of manipulating a small number of classes rather than the large number of observations against the slight loss of information caused by the grouping process.

Frequency distribution tables contain column headings of midpoints, frequencies, relative frequencies, cumulative frequencies, and relative cumulative frequencies. Actual observations form empirical distributions, as differentiated from the theoretical distributions studied in later chapters.

Graphic Representation

Graphs and charts are helpful in presenting and understanding the nature of the distributions of data.

Histogram. Histograms are formed by rectangles marking off the class intervals on the horizontal axis with each having the heights of either the frequencies or relative frequencies of the particular class interval.

The Polygon. If the midpoints of the top of the rectangles in a histogram are connected, the connecting lines form a frequency polygon.

The Ogive. If cumulative frequencies are used and the

connected points lie at the upper limit of each class, then the connected lines form a cumulative frequency polygon called an ogive (ō'jĭv) graph.

Pie Chart. Classifications may have special recognizable attributes other than being numerical. Such qualitative categories are better represented by a pie chart. The total pie is divided into the number of categories present with each slice proportional in size to the category's relative frequency. Pictograms and input/output charts are also useful in the display of data.

Patterns. Distributions can be described as being symmetrical, positively skewed, or negatively skewed. A positively skewed distribution is skewed to the right with its right-hand tail longer than the left-hand tail.

KEY TERMS

Array	The arrangement of the observations in the order of their magnitude--sequenced in either ascending or descending order.
Bar chart	A chart representing the magnitude of factors either quantitative or qualitative in nature.
Cluster sampling	A sampling procedure in which the population is first sampled by selecting random groups called clusters. These clusters can be the overall sample or a subsample of each cluster can make up the overall sample.
Cumulative frequency	The total number of observations equal to or less than a specified value.
Disproportionate sampling	A stratified sampling procedure in which the number of units in each subsample is inversely related to the homogeneity of the units in the stratum.
Empirical distribution	Any distribution based on actual observations of the researcher.
Frequency distribution	The arrangement of data to express the frequency of occurrences of observations in each class.
Histogram	A graphical representation of a frequency pattern of a set of observations using rectangles whose heights equal the frequency of their specific class interval.
Ogive	A line graph constructed by connecting the points at the upper limits of the class at heights of the cumulative frequency of the classes.

Pie chart	A round chart divided into sectors representing qualitative categories rather than numerical classes. The sizes of the sectors are proportional to the frequencies of the various categories.
Polygon	A line graph constructed by connecting the midpoints of the top of the rectangles of a histogram.
Proportionate sampling	A stratified sampling procedure whereby the number of units in each subsample is proportional to the relative number of units in each stratum.
Random sample	From an infinite population, the outcome of one observation does not affect the probability that any other observation will be selected. From a finite population, at each observation every available potential observation has an equal probability of being selected.
Relative frequency	The proportion of the total observations falling within a specified class.
Sampling unit	An observation from a population.
Simple random sampling	The procedure of mixing the population up completely and randomly drawing a sampling unit. The same procedure can be accomplished by giving each potential observation a unique number and then using a random number generator (table of random numbers, computer program, etc.) to select the next observation drawn.
Skewed distribution	A description of a pattern of a distribution. If the right-hand (positive) side of a distribution has a longer tail than the left, it is referred to as skewed to the right. If the tail on the negative side is longer, then the distribution is skewed to the left.
Strata	Classes or groups of relatively homogeneous units from a population.
Stratified sampling	A sampling procedure in which the population is divided into strata and subsamples are chosen from the strata and using simple random sampling procedures.
Systematic sampling	A sampling procedure whereby every kth unit of the population is chosen as a sampling unit. k is calculated as the size of the sample and is called the sampling ratio.

Two-stage sampling   A cluster sampling procedure in which subsamples of the clusters make up the overall sample.

ORGANIZED LEARNING QUIZ

1. To make inferences about a population, it is essential that any sample drawn

    a. is serially numbered.
    b. is normally distributed.
    c. is drawn one at a time.
    d. is representative of the population.

2. When sampling from an infinite population, a sample is considered a random sample

    a. if each observation does not affect the probability that any other will be chosen.
    b. if the sample size is finite.
    c. if observations are chosen in such a way as to increase the probability that others will be chosen in the future.
    d. if all other means of sampling cannot be tried.

3. An example of an infinite population is

    a. the sales of a company's product last year.
    b. the houses in a city with aluminum siding and asphalt shingles.
    c. the potential observations when tossing a coin.
    d. the potential observations when tossing a coin n times.

4. If a large sample is desired, cluster sampling might be advisable because of

    a. its frequently lower cost.
    b. its accuracy over simple random sampling.
    c. the homogeneity of large samples.
    d. its independence of structured interviews.

5. An arrangement of data in order of their magnitude is called

    a. a cumulative frequency distribution.
    b. an ogive chart.
    c. an array.
    d. an organization chart.

6. The construction of frequency tables requires first the selection of

    a. probabilities.
    b. class intervals.
    c. relative frequencies.
    d. a histogram.

7. Empirical distributions are based on

   a. three assumptions.
   b. theoretical observations.
   c. equal probabilities.
   d. actual data.

8. A frequency polygon is constructed by lines connecting

   a. the upper limits of the rectangles of a bar chart.
   b. the corners of an ogive chart.
   c. the midpoints of the relative cumulative frequency rectangles in a histogram.
   d. the midpoints of the tops of the rectangles of a histogram.

9. Pie charts may be particularly useful

   a. when dealing with numerical data rather than qualitative data categories.
   b. when dealing with qualitative categories rather than numerical data.
   c. unless the sample size is over 100.
   d. unless the sample size is over 360.

10. Positively skewed distributions

    a. have two identical halves.
    b. have long right-hand tails.
    c. have only nonnegative values.
    d. increase without bound to the right.

11. Five out of a sample size of 75 are included in a specific category. On a pie chart the category would occupy a sector size of _____ degrees.

12. As a rule of thumb, the number of class intervals of frequency distributions should not be fewer than _____ nor more than _____.

13. Relative frequencies are computed by dividing the frequency of a class by _____.

14. A distribution based on actual observations is called an _____ distribution in contrast to a _____ distribution.

15. Arrays can be sequenced in _____ or _____ order.

16. Explain why the midpoints of the class intervals should have as many digits as the raw data.

    _____

    _____

    _____

17. Explain the conditions necessary to ensure a random sample from a finite population.

    _____
    _____
    _____

18. Explain the use of a sampling ratio in systematic sampling.

    _____
    _____
    _____

19. Explain the difference between a frequency distribution and a cumulative frequency distribution.

    _____
    _____
    _____

20. Explain the difference between a frequency polygon and an ogive chart.

    _____
    _____
    _____

ANSWERS TO ORGANIZED LEARNING QUIZ

1. d   2. a
3. c   4. a
5. c   6. b
7. d   8. d
9. b   10. b

11. 24

12. 6, 20

13. the total number of observations (n)

14. empirical, theoretical

15. ascending or descending

16. This ensures that the lower and upper limits of each class will have an extra digit 5 and none of the raw data will fall on these limits.

17. At each selection of a sampling unit, all remaining potential observations have an equally likely chance of being drawn

next even though this probability changes after every draw.

18. The sampling ratio is the population size divided by the sample size. The quotient or ratio is labeled k. Simple random sampling of the first k units of the population determines the first sampling unit of the overall sample. Every next kth unit in the population completes the overall sample.

19. A frequency distribution expresses the number of observations in each class interval. A cumulative frequency distribution expresses the number of observations equal to or less than the midpoint of each class interval, assuming all observations of every class lie on the class midpoint.

20. The frequency polygon depicts either a frequency distribution or a relative frequency distribution, whereas the ogive is a cumulative frequency polygon that depicts either a cumulative frequency distribution or a relative cumulative frequency distribution.

SAMPLE EXERCISE

Develop a pattern of presentation that will help make the age in months of the automobiles driven by the faculty of the School of Business more informative.

37	28	3	38	8	40	15	39	55	29	49	24
14	36	42	1	51	17	52	1	79	38	21	60
48	7	34	28	31	38	18	44	60	6	47	64
29	24	12	40	15	5	42	25	9	52	47	31
1	41	52	29	74	36	23	48	37	43	14	3
34	4	25	13	9	32	67	34	1	62	94	54
44	14	45	46	30	50	11	38	24	61	38	12
8	43	33	18	40	19	60	36	44	83	8	58
47	29	5	37	3	43	47	26	17	62	70	28
16	58	38	24	48	39	6	55	68	52	14	40

The first step is to arrange the raw data in an array so as to make the numbers easier to manipulate.

```
1  6 12 17 25 30 36 39 43 47 52 62
1  6 13 18 25 31 37 39 43 48 54 62
1  7 14 18 26 31 37 40 44 48 55 64
1  8 14 19 28 32 37 40 44 48 55 67
3  8 14 21 28 33 38 40 44 49 58 68
3  8 14 23 28 34 38 40 45 50 58 70
3  9 15 24 29 34 38 41 46 51 60 74
4  9 15 24 29 34 38 42 47 52 60 79
5 11 16 24 29 36 38 42 47 52 60 83
5 12 17 24 29 36 38 43 47 52 61 94
```

Using a class interval of 15 months, fill in the missing items in the following frequency table.

Interval	Midpoint x	FR(x)	CF(x)	RF(x)	RCF(x)
		28			
15.5-30.5					
				0.717	
45.5-60.5			109		
				0.067	
			119		
					1

Complete a histogram, a frequency polygon, and an ogive for the data.

CASE PROBLEM

William and Marx Theaters

William and Marx Theaters operate a chain of movie theaters in a medium-sized North Central city. During the daytime, when the theaters are closed, space in their parking lots is rented to nearby officeworkers on a monthly basis.

The downtown location, however, is prime property and the organization has hired an attendant to be caretaker during the hours of 7:00 A.M. until 5:30 P.M. Hourly and daily rates are charged for the use of this parking lot. There is some question concerning the rates to be charged and Nancy Lutz has been asked to make a study and give her boss, "Choc" Marx, her recommendation.

As part of her investigation, she would like to determine some idea of customer utilization of the parking facility under the present pricing structure. Each customer is given a ticket with arrival time printed on the face and departure time is also printed; so when leaving she can determine the length of time spent by cars on the lot.

The parking tickets are kept in boxes large enough to hold one month's worth of tickets apiece. The boxes are numbered 1 to 12, representing one year of history. Nancy did not like the idea of recording the contents of all the boxes and decided to utilize some sort of sampling plan. She estimated that there were about 1500 tickets in each box.

Nancy felt that a sample of 100 observations should give her enough information about the characteristics of the pattern of the lengths of time spent for each customer.

1. What type of sampling plan do you suggest for Nancy? Are there alternatives? Evaluate.

2. What are Nancy's next steps after she picks the 100 tickets?

3. What type of information will she be able to report when her investigation is completed?

3

DATA DESCRIPTION: SAMPLE STATISTICS

CHAPTER LEARNING OBJECTIVES

Students should be able to

- Develop three measures of central tendency of sample observations.

- Describe the properties of the mean, median, and mode.

- Compare the peculiarities, limitations, and usefulness of the mean, median, and mode.

- Interpret the meaning of the range of sample data, giving the advantages and disadvantages of such a measure.

- Interpret the meaning of the average deviation of sample data, giving the advantage and disadvantage of such a measure.

- Interpret the meaning of the standard deviation of sample data, giving the advantages and disadvantages of such a measure.

- Correct the measure of the variance of sample data so that the corrected value is an unbiased estimate of the variance of the population from which the sample data were drawn.

- Estimate the number of observations occurring within a given number of standard deviations from the mean of any distribution and, in particular, the normal distribution.

SYNOPSIS

Measures of Central Tendency

If the observations of a sample are thought to be points on a number line, then a typical observation might be a middle one in which all the others cluster around. Such a typical observation is useful in helping describe the total set. It becomes the "spokesperson" for the body of data. Measures of this point are called *measures of central tendency* or, more simply, *averages*. More than one measure of an average is desired, since some distributions and circumstances are better explained by using one type of average than another.

The most popular measure of central tendency is the *mean*. For ungrouped data, the method of computing the mean is the same for any set of data, including the population. Its dimensional units are the same as for an individual observation; however, because of the operation of division in its calculation, its value need not be equal to any one of the individual observations. Also, a mean of whole numbers may very well be fractional in value. Important properties of the mean are (1) it can be substituted for each observation and the total of observations is unchanged; (2) the sum of each observation's distance from the mean (its deviation) is zero; (3) it allows a linear transformation of scale; (4) the sum of the squared deviations of each observation from the mean is minimized; and (5) the sample mean is an unbiased estimate of the population mean. Caution is suggested when using the mean (or any other measure of average) in making conclusions about a body of data. One value at a great distance from the mean can have a large influence in moving the value of the mean toward itself. This situation is possible, since all observations have an equal weight or influence on the value of the mean.

The *median* has different properties than the mean and, in certain circumstances, is a preferred typical measure. Its value is not pulled in the direction of skewness in a distribution as much as the mean. It also has the property of being the value that minimizes the absolute deviation explained below. The median is what most people associate with the middle observation--one-half above and one-half below--without being concerned with distances above or below. It is sometimes referred to as the 50 percentile point, meaning 50% of the observations lie somewhere below it.

The *mode* is probably the least useful mathematically of the three measures of average. The mode is not affected by skewness and thus does not explain its presence. Its use stems from its ease of computation. It is fairly simple to identify the value of observations that occur most frequently. If a data set has more than one mode, then one should consider partitioning the data into unimodal sets, since the mean and median are usually more informative when there is only one mode. Of course, a unimodal, symmetric distribution (such as the normal curve) has equal mean, median, and mode; so the determination of one gives all three.

Measures of Dispersion

When faced with a summary description of a set of data, giving a typical value like the mean is helpful but not at all complete. In describing scores of a rifle match, one might say that the average shot was a bull's eye. More information is needed about the dispersion of shots or consistency of values before determining the sharpshooter awards. To be more complete in a description of data, two parameters are necessary--a measure of central tendency and a measure of variability.

The easiest and most common measure is the *range*. Unfortunately, its value does not lend itself to further mathematical manipulation or statistical inference. It also hides the pattern of dispersion, since all observations could be grouped at either end or in the middle without changing the measure of the range. Unusually high or low values that distort the range are often eliminated by using the *interquartile range*.

Variability is related more directly to the mean if it is measured in terms of the amount of deviation from the mean that the observations possess. The *average deviation* totals all the distances from the mean for the observations (ignoring direction) and divides by the number of observations. Even though the median gives a minimum figure for the average deviation, the mean is customarily used. A data set with a larger value of average deviation than a second data set will then have a more scattered appearance.

The average deviation (sometimes referred to as the Mean Absolute Deviation) is useful when comparing two or more sets but is not too descriptive for a lone set. For this purpose, the *standard deviation* is used. Its most important property is that it has statistical inference significance and can be used in making inferences about the parent population. The standard deviation is the square root of the variance, and the method of calculating the variance varies, depending on the nature of the data, whether it is grouped or ungrouped.

Uses of Standard Deviation

The standard deviation of sample data is expressed in the same dimensional units as the observations. For instance, if data consist of ounces of breakfast cereal contained in a package filled by an automated process, then the standard deviation measure of variability is also measured in ounces. No matter what shape the distribution of sample data takes, knowing the standard deviation gives information about the concentration of the data about the mean. Tchebycheff's theorem gives bounds on the percentage of data points that must lie within any number of standard deviations from the mean greater than one. For instance, 56% of the data points lie within 1.5 standard deviations on either side of the mean.

Knowing (or assuming) the distribution of the data set is normal gives closer tolerances to the description of dispersion. For any normally distributed body of data, 87% of the data points lie within 1.5 standard deviations on either side of the mean. Percentages for other values of standard deviations are readily available in statistical tables giving the area under the normal curve. Small sample data sets are never completely normally distributed (they form discrete distributions rather than a continuous set); therefore approximations of the percentage of data points lying within given standard deviations of the mean must be assumed.

KEY TERMS

Absolute deviation   The value of the deviation without considering the sign or direction.

Average deviation	The arithmetic mean of the absolute values of the deviations from the mean, designated AD.
Bimodal distribution	A distribution with two modes in the data.
Deviation	A term usually referred to as the distance an observation lies from the arithmetic mean of the set, computed as $(X_i - \bar{X})$, where X_i is the specific observation number and $\bar{X}$ is the mean.
Dispersion	The scattering attribute of a set of data points. Also known as the set's variability.
Mean	The sum of all observations divided by the number of observations made. The most important measure of central tendency for inferential purposes. Designated as $\bar{X}$ or X-bar.
Measures of central tendency	Central values that give a mental picture of a data set to help make inferences about the nature of the population.
Median	The value that falls in the middle of a group of observations arranged in order of magnitude. The 50 percentile, designated as MD.
Mode	The value or class that has the highest frequency in a set of observations.
Percentiles	Position values that divide a group of observations arranged in order of magnitude into a hundred equal parts. There are 99 percentiles.
Quartiles	Position values that divide a group of observations arranged in order of magnitude into four equal parts. There are three quartiles.
Range	A measure of dispersion of a data set calculated by subtracting the lowest observation value from the highest observation value.
Standard deviation	A measure of dispersion of a data set calculated by taking the positive square root of the variance. It is the most important dispersion measure because of its inferential ability to help determine the nature of the population.
Unbiased estimator	A sample statistic whose long-run average value (expected value) is equal to a population parameter.

Variance   The mean of the squared deviations.

Weighted mean   The arithmetic mean in which each value
                            is weighted (multiplied) according to its
                            importance in the overall group.

FORMULA REVIEW

(3-1)   Arithmetic mean
        Unorganized data $\bar{X} = \dfrac{\sum_{i=1}^{n} X_i}{n}$ (for n observations)

(3-2)   Arithmetic mean
        Organized data $\bar{X} = \dfrac{\sum_{j=1}^{m} X_j \, FR(X_j)}{m}$ (for m distinct values or classes)

        Median
        Unorganized data $MD = \left(\dfrac{n+1}{2}\right)$ th observation in an array for n, an odd number.

                            $MD =$ any value between the $\left(\dfrac{n}{2}\right)$ th and the $\left(\dfrac{n+1}{2}\right)$ th observation for an even n.

(3-5)   Median
        Organized data $MD = L_m + \dfrac{\dfrac{n+1}{2} - CF(x_{m-1})}{FR(x_m)} (w)$

        where: L_m = the lower limit of the class that contains the median (the median class)

                $CF(x_{m-1})$ = the cumulative frequency below the median class

                $FR(x_m)$ = the frequency of median class

                w = the length of the median class

(3-6)   Average
        deviation $AD = \dfrac{\sum_{i=1}^{n} |X_i - \bar{X}|}{n}$ (for n observations)

(3-7)
and   Variance of X
(3-8)   Unorganized data $\hat{S}^2 = \dfrac{\sum_{i=1}^{n} (X_i - \bar{X})^2}{n} = \dfrac{\sum X^2 - \bar{X} \sum X}{n} = \dfrac{\sum X^2}{n} - \bar{X}^2$

(3-10) Variance of X
        Organized data $\hat{S}^2 = \dfrac{\sum_{j=1}^{m} (X_j - \bar{X})^2 \, FR(X_j)}{n}$ (for m classes)

(3-8) Standard deviation of X $\hat{S} = \sqrt{\hat{S}^2}$

(3-11) and (3-12) Estimator of population variance $S^2 = \dfrac{\sum_{i=1}^{n}(X_i-\bar{X})^2}{n-1} = \dfrac{(n)\hat{S}^2}{(n-1)}$

(3-16) Tchebycheff's theorem $RF(|X-\bar{X}| \leq ks) \geq 1 - \dfrac{1}{k^2}$ $(k>1)$

where   RF = relative frequency

k = the number of standard deviations

s = the standard deviation value

ORGANIZED LEARNING QUIZ

   Daily microwave oven sales in a large department store during the last ten days, arranged in numerical sequence, were 3, 4, 6, 10, 10, 11, 12, 12, 12, 14.

1. The mean number of sales was

   a. .94.
   b. 9.4.
   c. 94.
   d. 9.

2. The median number of sales was

   a. 9.
   b. 10.
   c. 10.5.
   d. 11.

3. The mode for the number of microwave ovens sold was

   a. 9.
   b. 10.
   c. 11.
   d. 12.

4. The range of the data set is

   a. 9.
   b. 10.
   c. 11.
   d. 12.

   The examination scores of five statistics students were 87, 93, 74, 98, and 48.

5. The mean of the examination scores is

    a. 80.
    b. 85.
    c. 90.
    d. 95.

6. The median is

    a. 90.
    b. 89.
    c. 88.
    d. 87.

7. The mode is

    a. 88.
    b. 89.
    c. 90.
    d. none of the above.

8. The range is

    a. 100.
    b. 50.
    c. 5.
    d. 25.

9. The 80 percentile value is

    a. 98.
    b. 48.
    c. 87.
    d. 74.

10. The data set is

    a. symmetrical.
    b. negatively skewed.
    c. positively skewed.
    d. continuous.

11. The average deviation is based on the difference between each value in the data set and the _____ of the group.

12. The middle 90% of a group would lie between the _____ percentile and the _____ percentile.

13. The sample variance is an unbiased estimate of the population variance when the sum of the squared deviations are averaged by dividing by _____.

14. The root-mean-squared deviation is a descriptive name of the _____.

15. When averaging the squared deviations of sample observations by dividing by n, the variance tends to _____ the true population variance.

16. The second quartile has the same value as the _____.

17. For a normal distribution of data points, _____% of the points will lie between the mean and three standard deviations above the mean. (Use the normal rule.)

18. For *any* distribution of data points, at least _____% of the points will lie between the mean and three standard deviations above the mean. (Use Tchebycheff's theorem.)

19. To calculate the range of the middle 50% of a data set, one would need _____.

20. For statistical inferences of population parameters, the most useful sample statistics are the _____ and the _____ or the _____.

21. Give an example where the range of data set might be a misleading value to explain the dispersion of values of the observations.

   _____
   _____
   _____

22. Give an example where the mode might be a misleading value to explain the central value of a data set.

   _____
   _____
   _____

23. Why might the range of the middle 80% of a data set be more useful than the range of the whole set?

   _____
   _____
   _____

24. Explain why the variance is a more important and useful measure of dispersion than either the range or average deviation.

   _____
   _____
   _____

25. What are the cautions one must observe when using a mean to describe a data set?

_____
_____
_____

ANSWERS TO ORGANIZED LEARNING QUIZ

1. b   2. c
3. d   4. c
5. a   6. d
7. d   8. b
9. a   10. b
11. mean   12. 5th, 95th
13. n - 1   14. standard deviation
15. underestimate   16. median
17. 50   18. 44
19. the first and third   20. mean, variance, standard deviation
    quartiles

21. One example is where there is one observation at an extremely high or low level in relation to the rest of the data, which are tightly grouped.

22. When a distribution is highly skewed in either direction.

23. The range of the middle 80% eliminates the top 10% and the bottom 10%. These two groups are likely to contain non-typical observations caused by error of measurement or some other assignable reason. Eliminating them increases the chance of obtaining a statistic better able to represent the data group.

24. The sample variance provides an unbiased estimate of the population variance. It is also more mathematically suited for further analyses.

25. It can be misleading because it is very sensitive to untypically high or low values. Circumstances may occur when an average is a poor descriptor of a set in the first place. This situation could happen when the observations are in several clusters or groups of values.

SAMPLE EXERCISE

Two electric power companies in the same city are interested in comparing the weekly salaries of their upper-middle managers. Company G has 9 executives in the classification of interest, whereas Company L has 13 executives in the same category. After considerable exchange and negotiation, the following data were made available to both companies.

	Company	
Person No.	G	L
1	$400	$420
2	400	420
3	430	420
4	450	430
5	500	430
6	580	500
7	580	500
8	580	520
9	670	520
10		520
11		520
12		600
13		700

When doing the hand computations (or even calculator computations) of means and standard deviations, a computational table is convenient to keep track of the important totals and subtotals. The basic table for a single data set with its explanations is given below. This table and its extensions will be used throughout this study guide.

	Company G	
Person No.	X	X^2
1	400	160,000
2	400	160,000
3	430	184,900
4	450	202,500
5	500	250,000
6	580	336,400
7	580	336,400
8	580	336,400
9	670	448,900
Totals	4,590 (A)	2,415,500 (B)
Means and correction	510 (C)	2,340,900 (D)
Variation		74,600 (E)
Variance		9,325 (F)
Standard deviation		$96.57 (G)

28

Explanation of Table

(A) Sum of observations $\sum_{i=1}^{n} X_i = 4{,}590$

(B) Sum of squares $\sum_{i=1}^{n} X_i^2 = 2{,}415{,}500$

(C) Arithmetic mean $\dfrac{\Sigma X}{n} = \dfrac{\text{\textcircled{A}}}{n} = \dfrac{4{,}590}{9} = 510$

(D) Correction factor necessary to reduce the sum of squares of each observation to the sum of squares of the deviations of each observation from the mean

$(\bar{X}) * (\Sigma X) = \text{\textcircled{C}} * \text{\textcircled{A}} = 510 * 4{,}590 = 2{,}340{,}900$

(E) Sum of squares of the deviations (Commonly called the variance) $\sum_{i=1}^{n}(X_i - \bar{X})^2 = \text{\textcircled{B}} - \text{\textcircled{D}} = 2{,}415{,}500 - 2{,}340{,}900 = 74{,}600$

(F) Sample variance $\dfrac{\Sigma(X-\bar{X})^2}{n-1} = \dfrac{\text{\textcircled{E}}}{n-1} = \dfrac{74{,}600}{8} = 9{,}325$

(G) Sample standard variation $\sqrt{\text{\textcircled{F}}} = \sqrt{9{,}325} = 96.57$

A computer program is given at the end of this chapter and has been run for both companies. Inspection of the computer output shows that although Company G has a slightly higher average salary figure, the salaries possess a higher measure of dispersion as measured by the standard deviation. The student is urged to study these two outputs and justify the tabular results. Also, notice that all observations lie within three standard deviations of the mean.

CASE PROBLEM

J. T. Wood Products Company

The cabinet division of a large wood products firm, J. T. Wood Company, Inc., makes high-quality solid-oak cabinets for kitchen and utility room use in private homes. They have six basic design styles that can be mixed in various ways to create a custom cabinet effect on customer demand. Ms. Joyce Tryhard, production control analyst, has been asked to report on the overall mean percentage of production productivity of cabinets for the last 2 months. Her immediate supervisor suspects that overall productivity has declined. Production figures for each style of cabinet in terms of percentage of full capacity for May and June are given below.

| | Productivity as a Percentage of Full Capacity | |
Cabinet Style	May	June
Heritage	95%	97%
Contemporary	100	90
Mediterranean	95	85
Colonial	92	98
Traditional	90	100
Oriental	60	90

The share of production of each style of cabinet has also changed each month and Ms. Tryhard was able to discover the following.

| | Total Production Share | |
Cabinet Style	May	June
Heritage	8%	10%
Contemporary	30	34
Mediterranean	35	38
Colonial	10	9
Traditional	9	6
Oriental	8	3
Totals	100%	100%

If Ms. Tryhard averaged the cabinet productivity figures for May and June, using an unweighted arithmetic mean, she would use the following calculations and would conclude that the average productivity rose from May to June.

	May	June
	95	97
	100	90
	95	85
	92	98
	90	100
	60	90
Totals	532	560
Means	$88\frac{2}{3}\%$	$93\frac{1}{3}\%$

However, a weighted average must be used, since all cabinets did not have an equal share of production and should not have equal weights in the averaging process. The production share is a convenient weighting criteria, for all weights will sum to one in both months. Ms. Tryhard's calculations now should look like this.

	Weighted Productivity	
Cabinet Style	May	June
Heritage	95% * .08 = 7.60	97% * .10 = 9.70
Contemporary	100% * .30 = 30.00	90% * .34 = 30.60
Mediterranean	95% * .35 = 33.25	85% * .38 = 32.30
Colonial	92% * .10 = 9.20	98% * .09 = 8.82
Traditional	90% * .09 = 8.10	100% * .06 = 6.00
Oriental	60% * .08 = 4.80	90% * .03 = 2.70
Totals	92.95%	90.12%

Therefore the overall average productivity did slip almost 3 percentage points in the 2 month period.

Computer Program

The BASIC program given will print the computation table for sample sums of squares, mean, variance, and standard deviation. The dimension statements in line 210 limit the program to 50 data points but can be changed by the user to accommodate any number. Outputs for the Sample Exercise demonstrate its use.

```
10 REM************************************************************
20 REM
30 REM       TITLE:          STATABLE
40 REM       LANGAUGE:       BASIC
50 REM       DESCRIPTIONS:
60 REM          THIS PROGRAM WILL PRINT THE COMPUTATION TABLE
70 REM          FOR SAMPLE SUMS OF SQUARES, MEAN, VARIANCE, AND
80 REM          STANDARD DEVIATION.
90 REM       INSTRUCTIONS:
100 REM         ENTER DATA AS REQUESTED BY THE PROGRAM.
110 REM      PROGRAMMER:
120 REM         GROVER WM. RODICH, PORTLAND STATE UNIV.
130 REM
140 REM************************************************************
150 REM
160 KEY OFF
170 CLS
180 PRINT "PROGRAM TO COMPUTE SUMS OF SQUARES, MEAN, VARIANCE,"
190 PRINT "AND STANDARD DEVIATION"
200 PRINT
210 DIM A(50),B(50)
220 PRINT "PLEASE INPUT THE NUMBER OF OBSERVATIONS IN THE DATA SET"
230 PRINT "N=";
240 INPUT N
250 PRINT "INPUT THE DATA ONE AT A TIME FOLLOWED BY A RETURN"
260 PRINT "WHAT IS THE FIRST NUMBER";
270 INPUT A(1)
280 PRINT "CONTINUE IN THE SAME FASHION"
290 FOR I=2 TO N
300 INPUT A(I)
310 NEXT I
320 PRINT
330 PRINT
340 PRINT
350 PRINT
360 PRINT "ITEM #          X            X-SQUARE"
370 PRINT
380 FOR I=1 TO N
390 PRINT I,A(I),A(I)*A(I)
400 NEXT I
410 PRINT "-------------------------------------------"
420 PRINT
430 C=0
440 D=0
450 FOR I=1 TO N
```

```
460 C=C+A(I)
470 D=D+A(I)*A(I)
480 NEXT I
490 PRINT "TOTALS",C,D
500 E=C/N
510 F=C*E
520 PRINT
530 PRINT "MEANS AND",E,F
540 PRINT "CORRECTION","         ","----------------"
550 PRINT
560 PRINT"VARIATION","         ",D-F
570 PRINT
580 PRINT"VARIANCE","       ",(D-F)/(N-1)
590 PRINT
600 PRINT"ST. DEVIATION","        ",((D-F)/(N-1))^.5
610 PRINT
620 PRINT
630 PRINT
640 KEY ON:END
```

```
PROGRAM TO COMPUTE SUMS OF SQUARES, MEAN, VARIANCE,
AND STANDARD DEVIATION

PLEASE INPUT THE NUMBER OF OBSERVATIONS IN THE DATA SET
N=? 9
INPUT THE DATA ONE AT A TIME FOLLOWED BY A RETURN
WHAT IS THE FIRST NUMBER? 400
CONTINUE IN THE SAME FASHION
? 400
? 430
? 450
? 500
? 580
? 580
? 580
? 670
```

ITEM #	X	X-SQUARE
1	400	160000
2	400	160000
3	430	184900
4	450	202500
5	500	250000
6	580	336400
7	580	336400
8	580	336400
9	670	448900
-----------	-----------	-----------
TOTALS	4590	2415500
MEANS AND CORRECTION	510	2340900 ----------------
VARIATION		74600
VARIANCE		9325
ST. DEVIATION		96.56603

```
PROGRAM TO COMPUTE SUMS OF SQUARES, MEAN, VARIANCE,
AND STANDARD DEVIATION

PLEASE INPUT THE NUMBER OF OBSERVATIONS IN THE DATA SET
N=? 13
INPUT THE DATA ONE AT A TIME FOLLOWED BY A RETURN
WHAT IS THE FIRST NUMBER? 420
CONTINUE IN THE SAME FASHION
? 420
? 420
? 430
? 430
? 500
? 500
? 520
? 520
? 520
? 520
? 600
? 700
```

ITEM #	X	X-SQUARE
1	420	176400
2	420	176400
3	420	176400
4	430	184900
5	430	184900
6	500	250000
7	500	250000
8	520	270400
9	520	270400
10	520	270400
11	520	270400
12	600	360000
13	700	490000
TOTALS	6500	3330600
MEANS AND CORRECTION	500	3250000
VARIATION		80600
VARIANCE		6716.667
ST. DEVIATION		81.95527

4

EVENTS AND PROBABILITIES

CHAPTER LEARNING OBJECTIVES

Upon completion of this chapter the student should be able to

- Provide definitions for the key terms.
- Differentiate between a permutation and combination.
- Calculate the number of outcomes for simple and compound events.
- Calculate probability by use of counting rules.

SYNOPSIS

This chapter introduces the basic terminology associated with the area of study called probability and the methods for computing that probability.

The area of probability is important because it serves as the foundation for inferential statistics. Inferential statistics probably constitutes 60 to 70% of a typical business statistics course. Mastering the material in inferential statistics depends heavily on an understanding of probability. In addition, probability serves as an important tool of analysis in business decisions in its use in the area of expected value and decision theory.

Outcome and Random Events

The simplest notion of probability is that it is the proportion of times that a particular event will occur. Thus we make statements that a head will occur 50% of the time on the flip of a coin, a heart will occur 25% of the time when we draw a card from a playing deck, or an even number will occur 50% of the time when we roll a die. All are statements of probability and reflect the percentage of time that we believe each event will occur. The approach to computing this probability is to count the outcomes of an experiment. We count the number of ways that we can realize the event of interest (often termed the number of successes) and we count the total number of possible outcomes to our experiment. The probability of the event is then the number of successes divided by the total number or outcomes.

$$P(A) = \frac{\text{number of ways A occurs}}{\text{total number of outcomes}}$$

Since counting is essential to calculating probability, it is important to understand the methods of counting. The simplest technique is complete enumeration. We simply write down each possible outcome of the experiment. Next we count the total number of outcomes; and we can count those outcomes that meet our criteria for success and compute our probability by dividing the number of successes by the total number of outcomes.

Rules for Counting Events

In problems where the number of outcomes is small, complete enumeration is a simple and expedient method of counting. Many problems, though, involve possible outcomes that number in the thousands, millions, and even higher. Complete enumeration is not possible under such circumstances. Under these conditions we must rely on mathematical formulas or rules to enable us to calculate the number of outcomes.

The first rule of importance in counting outcomes is the multiplication rule for counting events. If we have n acts that can occur in k_1, k_2, k_3, ..., k_n ways and if the outcome of any act in no way influences the number of possible outcomes of any other act, then we can calculate the total number of possible outcomes by multiplying:

$$(k_1)(k_2) \cdots (k_n).$$

The next rule for counting involves permutations. In this situation, we are to select r things from n things. For example, we may select two (r) cards from a deck of 52 (n) playing cards. If the order of cards selected is important (i.e., selecting an ace and then a king is not the same as selecting a king and then an ace), then the number of possible outcomes can be calculated by

$$P_r^n = n(n - 1)(n - 2) \cdots (n - r + 1).$$

The last rule involves selecting r things from n things where order is not important (i.e., selecting an ace and then a king is considered the same as selecting a king and then an ace). This is termed the number of combinations and can be calculated by

$$C_r^n = \frac{n(n - 1)(n - 2) \cdots (n - r + 1)}{r!}.$$

The student should be cautioned that the application of these counting rules depends on the assumption that all outcomes are equiprobable.

KEY TERMS

Combination	A combination is a subset of r objects from a set of n objects. Combinations differ from permutations in that the latter consider the order in which objects are selected and combinations do not.
Compound event	A compound event is a subset of the sample space and it is divisible into two or more simple random events.
Experiment	An experiment is some process or operation that leads to well-defined outcomes.
Outcome	An outcome is the result of a single trial of an experiment.
Permutation	A permutation is an ordered arrangement of r objects from n objects.
Probability of an event	The probability of an event is the proportion of time that the event will occur in the long run.
Random event	A random event is an event in which outcomes are not predictable with certainty.
Sample space	The sample space of an experiment is the set of all possible distinct outcomes of the experiment.
Simple event	A simple random event is the outcome of a single trial in any particular experiment.
Trial	A trial is an act that leads to one of the possible distinct outcomes of the experiment.

FORMULA REVIEW

(4-1) Multiplication rule

If a number (n) of acts are carried out, and each can be performed in the same number of ways (k), then the total number of possible outcomes for n acts is

$$(k)(k) \cdots (k) \text{ or } k^n.$$

(4-2) Multiplication rule

If there are n acts that can be performed in $k_1, k_2 \cdots k_n$ ways, respectively, then the total number of different possible outcomes for the n acts in succession is

$$(k_1)(k_2) \cdots (k_n).$$

(4-3) Permutations   The number of permutations of r objects taken from a set of n objects is

$$P_r^n = n(n-1)(n-2) \cdots (n-r+1)$$

where $(n > r)$

(4-4) Permutations $P_r^n = \dfrac{n!}{(n-r)!}$.

(4-5) Combinations   The number of combinations of r objects taken from a set of n objects is

$$C_r^n = \dfrac{n!}{(n-r)!r!} = \dfrac{n(n-1)(n-2)\cdots(n-r+1)}{r!}$$

(4-6) Probability $P(A) = \dfrac{n}{N} = \dfrac{\text{the number of possible outcomes in A}}{\text{the total number of possible outcomes in the sample space}}$

ORGANIZED LEARNING QUIZ

1. If we throw a die and get a 6, the 6 represents one _____ of the experiment.

2. The question of whether to use the formula for a permutation or combination involves whether _____ is important or not.

3. The value of 0! is _____.

4. In order to calculate probability using counting rules, our events must be _____.

5. Probability is the _____ of times a particular event will occur.

6. Counting rules are useful because they can be used in calculating _____.

7. I will draw three cards from an ordinary playing deck of 52 cards. After each draw I will replace the card. I am interested in the event drawing a 2, 3, 4, in order. I should use the *permutation/combination* rule to count these events.

8. The sample space is the set of all possible _____.

9. In calculating probability, the total number of outcomes goes in the *numerator/denominator*, whereas the number of successes goes in the *numerator/denominator*.

10. Drawing an ace or king from a deck of cards is an example of a *simple/compound* event.

11. Compute the following

    a. 4!

    b. 6!

    c. C_3^5

    d. P_2^6

Ten candidates are interviewed for a job. You are asked to select the top five candidates. Suppose that, unfortunately, your selection process is completely random.

12. a. How many ways can you select five candidates? (Order is not important.) _____

     b. What is the probability of selecting the five candidates that are the best qualified? _____

13. a. How many ways can you select the five candidates in order? _____

     b. What is the probability of selecting the five best-qualified candidates in the right order? _____

14. You will draw three cards from an ordinary deck of cards.

     a. How many ways can you draw three cards? (Order is not important.) _____

     b. How many ways can you draw three aces? _____

     c. What is the probability of drawing three consecutive aces? _____

15. Repeat problem 14, but replace each card after it is drawn. (Each of the three draws is from a deck of 52 cards.)

     a. _____

     b. _____

     c. _____

16. A traveling salesman has 20 accounts in different cities. He must select eight cities to visit.

     a. How many combinations of eight cities is possible? _____

     b. How many different sequences of eight cities is possible if he must also select the order in which he will visit the cities? _____

17. Refer to problem 16. If he divides the cities into two groups of ten, what is the probability that he will visit 2 cities in the first group of ten and 6 cities in the second group of ten if he selects his 8 cities at random? _____

18. A shipment of 20 parts has just arrived. There are two defective items in the 20. You take a sample of four parts.

    a. What is the probability that you will get at least one bad part in the sample of four? _____

    b. What is the probability that you will get both bad parts? _____

19. A company has 3 manufacturing plants and 12 warehouses. Plant A can supply 6 warehouses, plant B can supply 4 warehouses, and plant C can supply 2 warehouses. A warehouse is only supplied from a single plant.

    a. How many shipping plans are possible? _____

    b. If there is a single best plan, what is the probability of finding this plan by chance? _____

20. A dietician must plan a menu from the following selections.

Meat	Vegetables
Ham	Peas
Steak	Carrots
Hamburger	Beans
Chicken	Potatoes
	Broccoli

    The dietician will select one meat and two vegetables.

    a. How many menus are there? _____

    b. You dislike chicken, beans, carrots, and broccoli. What is the probability that you will get a menu where you dislike at least one item? _____

21. How can one distinguish between a simple and compound event?
    _____
    _____
    _____

22. A random event is typified by the fact that we cannot predict the event. Do you agree or disagree with this statement and why?
    _____
    _____
    _____
    _____

23. From a business statistics point of view, why is probability an important topic?

   _____
   _____
   _____
   _____

24. Why can't we use complete enumeration in place of complicated counting formulas?

   _____
   _____
   _____
   _____

25. Distinguish between when you should use the multiplication rule, the rule for permutations, or the rule for combinations.

   _____
   _____
   _____
   _____

ANSWERS TO ORGANIZED LEARNING QUIZ

1. outcome
2. order
3. 1
4. equiprobable
5. proportion
6. probability
7. permutation
8. events
9. denominator, numerator
10. compound

11. a. $4! = 4 \cdot 3 \cdot 2 \cdot 1 = 24$

    b. $6! = 6 \cdot 5 \cdot 4 \cdot 3 \cdot 2 \cdot 1 = 720$

    c. $C_3^5 = \dfrac{5 \cdot 4 \cdot 3}{3 \cdot 2 \cdot 1} = 10$

    d. $P_2^6 = 6 \cdot 5 \cdot 4 \cdot 3 = 360$

12. a. $C_5^{10} = \dfrac{10 \cdot 9 \cdot 8 \cdot 7 \cdot 6}{5 \cdot 4 \cdot 3 \cdot 2 \cdot 1} = 252$

    b. $P(A) = \dfrac{1}{252}$

13. a. $P_5^{10} = 10 \cdot 9 \cdot 8 \cdot 7 \cdot 6 = 30,240$

13. b. $P(A) = \dfrac{1}{30,240}$

14. a. $C_3^{52} = \dfrac{52 \cdot 51 \cdot 50}{3 \cdot 2 \cdot 1} = 22,100$

    b. $C_3^4 = \dfrac{4 \cdot 3 \cdot 2}{3 \cdot 2 \cdot 1} = 4$

    c. $P(A) = \dfrac{4}{22,100}$

15. a. $52 \cdot 52 \cdot 52 = 140,608$

    b. $4 \cdot 4 \cdot 4 = 64$

    c. $P(A) = \dfrac{64}{140,608}$

16. a. $C_8^{20} = \dfrac{20 \cdot 19 \cdot 18 \cdot 17 \cdot 16 \cdot 15 \cdot 14 \cdot 13}{8 \cdot 7 \cdot 6 \cdot 5 \cdot 4 \cdot 3 \cdot 2 \cdot 1} = 125,970$

    b. $P_8^{20} = 20 \cdot 19 \cdot 18 \cdot 17 \cdot 16 \cdot 15 \cdot 14 \cdot 13 = 5,079,110,400$

17. $P(A) = \dfrac{(C_2^{10})(C_6^{10})}{C_8^{20}} = \dfrac{\dfrac{(10 \cdot 9)}{(2 \cdot 1)} \dfrac{(10 \cdot 9 \cdot 8 \cdot 7 \cdot 6 \cdot 5)}{(6 \cdot 5 \cdot 4 \cdot 3 \cdot 2 \cdot 1)}}{125,970} = \dfrac{(45)(1260)}{125,970} = \dfrac{56,700}{125,970}$

18. a. P(at least 1 bad part) = 1 - P(no bad parts) = $1 - \dfrac{C_4^{18}}{C_4^{20}}$

    $= 1 - \dfrac{\dfrac{(18 \cdot 17 \cdot 16 \cdot 15)}{4 \cdot 3 \cdot 2 \cdot 1}}{\dfrac{(20 \cdot 19 \cdot 18 \cdot 17)}{4 \cdot 3 \cdot 2 \cdot 1}} = 1 - \dfrac{3060}{4845} = 1 - .63$

    $= .34$

    b. P(2 bad parts) $= \dfrac{C_2^2 C_2^{18}}{C_4^{20}} = \dfrac{(2)\dfrac{(18 \cdot 17)}{(2 \cdot 1)}}{4845} = \dfrac{153}{4845} = .03$

19. a. $6 \cdot 4 \cdot 2 = 48$

19. b. $P(A) = \dfrac{1}{48}$

20. a. $C_1^4 - C_2^5 = 4 - \dfrac{5-4}{2-1} = 40$

   b. $P(\text{menu you dislike}) = 1 - P(\text{menu you like}) = 1 - \dfrac{C_1^3 \cdot C_2^2}{C_1^4 \cdot C_2^5}$

$$= 1 - \dfrac{3 \cdot 1}{4 \dfrac{(5 \cdot 4)}{(2 \cdot 1)}} = \dfrac{37}{40}$$

21. An event is a compound event if it can be divided into two or more other simple events. It is a simple event if it cannot be further subdivided.

22. A random event is typified by the fact that we do not know the outcome with certainty. We may be able to predict the event with a probabilistic statement, such as we believe that there is a 50% chance of flipping a head. Here we have made a prediction; however, our prediction is not based on certainty.

23. It is important because it serves as the underlying theory in inferential statistics. It is also important in calculating expected value and in decision theory.

24. The numbers become so large that it is almost impossible to list all possible events.

25. The multiplication rule can be applied when we have events in sequence. The total number of outcomes is the product of the number of times each event can occur for all events. The permutation rule deals with selecting r objects from n objects when order is important. The combination rule deals with selecting r objects from n objects when order is not important.

SAMPLE EXERCISE

   The Mellow Beer Company produces and distributes beer. They have 3 factories and 18 distributors. Each factory and warehouse produces or carries a full line of products. The products are as follows.

Products	Containers	Packaged
Light	12-oz. cans	6 pack
Regular	12-oz. bottles	8 pack
Premium	8-oz. cans	
	8-oz. bottles	

1. In order to carry a full inventory, how many different products/containers/packages must a distributor carry?

   _____

2. Production setups depend on the type of product and whether it is cans or bottles. How many different production setups are possible? _____

3. If orders for product/containers/packages are all equally likely, what is the probability of receiving an order for light beer? (Type of containers and package is not important.) _____

ANSWERS TO SAMPLE EXERCISE

1. $3 \cdot 4 \cdot 2 = 24$

2. $3 \cdot 2 = 6$

3. $P(A) = \dfrac{C_1^3 C_1^4 C_1^1}{C_1^3 C_1^4 C_1^2} = \dfrac{3 \cdot 4 \cdot 1}{3 \cdot 4 \cdot 2} = \dfrac{1}{2}$

CASE PROBLEM

A shipment of 20 items has just been received and must be subjected to acceptance inspection. We will select three items and test. (The test is a destructive test.) If the first item is bad, we reject the remainder of the 20 items. All other conditions lead to acceptance. Of the 20 items, four are bad.

a. What is the probability of rejecting the lot on the first item?

b. What is the probability of rejecting the lot if the first item was good?

c. What is the probability of accepting the lot?

COMPUTER PROGRAM

The BASIC program given will calculate combinations and permutations and will print out the calculations. All input is prompted during execution of the program. A sample run is provided.

```
10 REM***************************************************************
20 REM
30 REM        TITLE:       PERM-COM
40 REM        LANGAUGE:    BASIC
50 REM        DESCRIPTIONS:
60 REM           THIS PROGRAM WILL CALCULATE COMBINATIONS AND
70 REM           PERMUTATIONS.
80 REM        INSTRUCTIONS:
90 REM           ENTER DATA AS REQUESTED BY THE PROGRAM.
100 REM
110 REM***************************************************************
120 REM
130 KEY OFF
140 CLS
150 PRINT "PROGRAM TO CALCULATE PERMUTATIONS AND COMBINATIONS"
160 PRINT
170 PRINT
180 PRINT "INPUT AS FOLLOWS"
190 PRINT "TYPE A 2 IF CALCULATING PERMUTATIONS"
200 PRINT "TYPE A 1 IF CALCULATING COMBINATIONS"
```

```
210 PRINT "TYPE A 0 TO TERMINATE THE PROGRAM";
220 INPUT K
230 R1=1
240 IF K=0 THEN 780
250 IF K>2 THEN 180
260 IF K<0 THEN 180
270 PRINT
280 PRINT
290 PRINT "PROBLEM NUMBER";
300 INPUT N
310 PRINT "N = ";
320 INPUT N
330 IF N <=0 THEN 310
340 PRINT "R = ";
350 INPUT R
360 IF R<1 THEN 340
370 IF R>N THEN 340
380 PRINT "CALCULATE N(N-1)(N-2)....(N-R+1)"
390 N1=1
400 FOR I=N TO N-R+1 STEP -1
410 N1=N1*I
420 IF I=N-R+1 THEN 450
430 PRINT I;"   X   ";
440 NEXT I
450 PRINT I;"   =   ";N1
460 A=N1
470 REM CALCULATE COMBINATIONS IF K=1
480 PRINT
490 PRINT
500 IF K=2 THEN 580
510 GOSUB 650
520 PRINT
530 PRINT
540 A=N1/R1
550 REM OUTPUT ANSWER
560 PRINT "NUMBER OF COMBINATIONS OF "
570 GOTO 590
580 PRINT "NUMBER OF PERMUTATIONS OF"
590 PRINT "R THINGS FROM N =";N1;"/";R1;"=";N1/R1
600 PRINT
610 PRINT
620 LOCATE 24,1:PRINT "TO CONTINUE PRESS ANY KEY";:A$=INPUT$(1)
630 GOTO 140
640 STOP
650 PRINT "CALCULATE R FACTORIAL"
660 R1=1
670 IF R>1 THEN 700
680 PRINT "R! = 1"
690 RETURN
700 PRINT R1;"   X   ";
710 FOR I=2 TO R
720 R1=R1*I
730 IF I=R THEN 760
740 PRINT I;"   X   ";
750 NEXT I
760 PRINT I;"   =   ";R1
770 RETURN
780 KEY ON:END
```

```
PROGRAM TO CALCULATE PERMUTATIONS AND COMBINATIONS

INPUT AS FOLLOWS
TYPE A 2 IF CALCULATING PERMUTATIONS
TYPE A 1 IF CALCULATING COMBINATIONS
TYPE A 0 TO TERMINATE THE PROGRAM? 1

PROBLEM NUMBER? 1
N = ? 20
R = ? 4
CALCULATE N(N-1)(N-2)....(N-R+1)
 20    X    19   X    18   X    17   =    116280


CALCULATE R FACTORIAL
  1   X    2   X    3   X    4   =    24


NUMBER OF COMBINATIONS OF
R THINGS FROM N = 116280 / 24 = 4845

TO CONTINUE PRESS ANY KEY


PROGRAM TO CALCULATE PERMUTATIONS AND COMBINATIONS

INPUT AS FOLLOWS
TYPE A 2 IF CALCULATING PERMUTATIONS
TYPE A 1 IF CALCULATING COMBINATIONS
TYPE A 0 TO TERMINATE THE PROGRAM? 2

PROBLEM NUMBER? 2
N = ? 20
R = ? 3
CALCULATE N(N-1)(N-2)....(N-R+1)
 20    X    19   X    18   =   6840


NUMBER OF PERMUTATIONS OF
R THINGS FROM N = 6840 / 1 = 6840



TO CONTINUE PRESS ANY KEY

PROGRAM TO CALCULATE PERMUTATIONS AND COMBINATIONS

INPUT AS FOLLOWS
TYPE A 2 IF CALCULATING PERMUTATIONS
TYPE A 1 IF CALCULATING COMBINATIONS
TYPE A 0 TO TERMINATE THE PROGRAM? 0
Ok
```

5

PROBABILITY RULES AND FUNCTIONS

CHAPTER LEARNING OBJECTIVES

Upon completion of this chapter, the student should be able to

- Provide definitions for the key terms.
- Define and apply the concept of mutually exclusive events, independent events, and dependent events.
- Be able to apply the addition rule.
- Be able to apply the multiplication rule.
- Define the term random variable and be able to indicate the random variable in problem situations.
- Define the term probability function.
- Define the term probability distribution.

SYNOPSIS

This chapter extends the ways for computing probability through the use of the addition and multiplication laws. The concepts of a random variable and a probability distribution are also introduced.

Addition Rules

Up to now the method for computing probability has consisted of counting successes and dividing by the total number of outcomes. It required that each event be equally likely. An additional way of computing probabilities involving compound events is the use of the addition or the multiplication law. The addition and multiplication laws extend our abilities to compute probabilities to more general cases.

The addition rule covers the probability of two events in which success is represented by the occurrence of either event or both.

The only way not to have a success is when neither event occurs. For example if we wish to know the probability of an ace or a heart on the draw of one card, the addition law will apply. Success is represented by the draw of an ace or the draw of a heart or the draw of the ace of hearts. Failure is represented by drawing a card other than an ace or a heart. The student may find it easier to know when to apply the addition rule by noting that the word connecting the two events is "or." Thus we say success is one event *or* the other event occurring. The "or" relationship is the key to knowing that the addition law should be applied.

The multiplication law covers the probability of two events in which success is represented by both events occurring. If we were interested in the probability of drawing an ace and a heart, then success would only be represented by the draw of the ace of hearts. Drawing an ace other than a heart would be a failure, drawing a heart other than the ace of hearts would be a failure, and drawing a card other than an ace or a heart would be a failure. The clue to identifying when the multiplication rule should be applied is the use of the word "and" to connect the two events.

The key to determining which rule to use is to state the problem in such a way as to use either the connecting phrase "or" or "and" for the two events. One or the other will fit the problem. if it is an "or," apply the addition rule; if it is an "and," apply the multiplication rule. In set theory the union symbol $\cup$ is equivalent to "or" and the intersection symbol $\cap$ is equivalent to "and."

The addition law in its general form is

$$P(A \text{ or } B) = P(A) + P(B) - P(A \text{ and } B).$$

There is a special case of this formula when A and B are mutually exclusive events. In mutually exclusive events, if one event occurs, the other cannot. Drawing an ace or a king represent mutually exclusive events. If an ace is drawn, then a king cannot be drawn. Hence $P(A \text{ and } B) = 0$, and the special form of the addition law when A and B are mutually exclusive events is

$$P(A \text{ or } B) = P(A) + P(B).$$

The multiplication rule in its general form is

$$P(A \text{ and } B) = P(A) \cdot P(B|A).$$

$P(B|A)$ is a conditional probability and is interpreted as the probability of B, given that we know that A has occurred.

Independent Events

Again, there is a special case of this rule when events A and B are independent. Independent events are events in which knowing that one event has occurred does not change the knowledge about the probability of the other event. If A and B are independent, then $P(B|A) = P(B)$. The special form of the multiplication law when A and B are independent is

$$P(A \text{ and } B) = P(A) \cdot P(B).$$

Dependent Events

A variation of the general multiplication rule provides the formula for conditional probability.

$$P(B|A) = \frac{P(A \text{ and } B)}{P(A)}$$

By stating that A has occurred, we have limited our sample space to joint events that include A. Success becomes the proportion of times that A and B occur together divided by the total sample space, the proportion of times that A occurs.

Probability Function

In most situations, it is convenient to assign numerical values to the outcomes of experiments. In some cases, this step may be arbitrary, such as in coin flipping, where we may assign a 1 to heads and a 0 to tails. In other cases, it may have a natural meaning, such as the number of spots on a die. The numerical value assigned to events is called a random variable. A discrete random variable is one that is obtained by counting and only takes on positive integer values. A continuous random variable is one that is obtained by measurement and can take on decimal or fractional values.

Closely associated with the ideas of a probability function is the idea of a probability distribution. A probability distribution is the identification of all possible distinct outcomes and their associated probability. This identification may be through tabulation, formula, or rule.

KEY TERMS

Axioms	Statements accepted without proof.
Conditional probability	Probability that an event will occur, given that another event has occurred.
Dependent events	Two events for which the occurrence of one affects the probability of the other occurring.
Functional relationship	Relationship between the probability of a variable and possible values of the variable.
Independent events	Two events for which the occurrence of one does not affect the probability of the other occurring.
Joint event	The intersection of two events.
Joint probability	Probability of a joint event.
Marginal probability	Total of the probabilities for any given column or row in a joint probability table.

Overlapping events		Events that have a portion of the sample space in common.
Probability distribution		Distribution of probabilities, each associated with one of the possible distinct values of the random variable.
Probability function		Rule that assigns a probability fraction to each of the distinct values of the random variable.
Probability rules		Multiplication rules for the calculation of probabilities.
Random variables		Given a variable made up of events $E_1, \ldots, E_n$, the variable is called random if the occurrence of the events is determined solely by chance.

FORMULA REVIEW

(5-1) Complement union $P(E) = 1 - P(E')$ and $P(E') = 1 - P(E)$

(5-2) Special addition rule: mutually exclusive events $P(E_1 \cup E_2) = P(E_1) + P(E_2)$

(5-3) General addition rule $P(E_1 \cup E_2) = P(E_1) + P(E_2) - P(E_1 \cap E_2)$

(5-4) Union: mutually exclusive events $P(E_1 \cup E_2 \cup \cdots \cup E_n) = P(E_1) + P(E_2) + \cdots + P(E_n)$

(5-5) Special multiplication rule: independent events $P(A \cap B) = P(A) \cdot P(B)$

(5-6) General multiplication rule $P(A \cap B) = P(A) \cdot P(B|A)$

(5-7) Conditional probability $P(B|A) = \dfrac{P(A \cap B)}{P(A)}$

ORGANIZED LEARNING QUIZ

1. True or False: Event E will only have a complement E' when the P(E) is less than 1.

    a. True
    b. False

2. Suppose that a farmer has six sows with litters. Sow A has 8 pigs, sow B 9 pigs, sow C 10 pigs, sow D 10 pigs, sow E 11 pigs, and sow F 12. Let E_1 be the event of selecting a litter at random containing less than 10 pigs. Let E_2 be the event

of selecting a litter containing more than 10 pigs. The probability of E_1 or E_2 ($E_1 \cup E_2$) is

a. .5.
b. 1.0.
c. neither of the above.

3. True or False: In problem 2 the probability of the complement of event E_1 is equal to the probability of event E_2.

a. True
b. False

4. $E_1 \cap E_2 =$

a. .6.
b. .67.
c. .0.

5. In problem 2 change the definition of events so that E_1 is the event of selecting a litter with less than 11 pigs and E_2 is the event of selecting a litter with more than 8 pigs. $P(E_1 \cup E_2) =$

a. 1.0.
b. 1/3.
c. 2/3.

6. The probability that you will get out of bed on time is .9; the probability that your car will start is .5; the probability that you will get out of bed on time *and* that your car will start is .7. What is the probability that you will get up on time *or* that your car will start?

a. 1.0.
b. .85.
c. .7.

7. Two normal dice are rolled. What is the probability that the numbers showing will total more than six *or* will be even?

a. 35/36.
b. 32/36.
c. 30/36.

8. Suppose that the probability that you will study statistics (S) this evening is .2; chemistry (C), .25; play cards (PC), .3; suppose that you will only do one of them. What is the probability that you will do one of these things?

a. 1.0.
b. .75.
c. .55.
d. .45.

9. Two fair coins are flipped twice in succession. What is the probability of the coins landing as two heads and then two tails?

50

a. .05.
   b. .004.
   c. .125.
   d. .0625.

10. Two eggs in a tray of one dozen are cracked. If two are drawn without replacement, what is the probability of picking both cracked eggs?

    a. 1/144.
    b. 1/132.
    c. 1/121.
    d. 2/144.

11. The probability of an event must be a _____ value or zero.

12. P(U) = _____. U stands for the _____.

13. P(E) must be (less than), (less than or equal to), (more than or equal to) 1.

14. In problem 2 E_1 and E_2 are said to be _____ requiring the use of the special addition rule.

15. Indicate which of the following are independent events and which are dependent.

    a. The consumption of alcoholic beverages by teenagers and the number of accidents involving teenage drivers.
    b. Getting a head on the fourth flip of a fair coin when the three prior flips had resulted in a head.
    c. Washing your car and having it rain.

16. The intersection of two events is called a _____ _____.

17. The probability of the intersection of two events is called a _____ _____.

18. The probability that an event will occur, *given* that another event has already occurred, is called a _____ _____.

19. A probability function shows the probability for each of a number of _____ _____ subsets. These probabilities add to _____.

20. A probability distribution is a _____ _____ in contrast to an empirical relative frequency distribution.

21. A variable is said to be random if _____ _____ _____.

51

22. Give examples of independent events.

_____
_____
_____

23. Give examples of dependent events.

_____
_____
_____

24. Explain the construction of a joint probability table.

_____
_____
_____

ANSWERS TO ORGANIZED LEARNING QUIZ

1. b. if $P(E) = 1$, $P(E') = 0$.

2. c. $P(E_1) = 1/3$, $P(E_2) = 1/3$

3. b. $P(E') = 1 - P(E_1) = 2/3$

4. c.

5. a. $P(E_2) = 5/6$; $p(E_1) = 2/3$; therefore $P(E_1 \cup E_2) = 1.0$

6. c. $P(U)$ = probability of getting out of bed on time
       $P(W)$ = probability of car starting
       $P(U \cup W) = P(U) + P(W) - P(U\ W) = .9 + .5 - .7$

7. c. $P(\text{Even}) = 1/36 + 3/36 + 5/36 + 5/36 + 3/36 + 1/36 = 18/36$
       $P(>6) = 6/36 + 5/36 + 4/36 + 3/36 + 2/36 + 1/36 = 21/36$
       $P(\text{Even and} >6) = P(8) + P(8) + P(10) + P(12)$
       $\qquad = 5/36 + 3/36 + 1/36 = 9/36$
       $P(\text{E or} >6) = 18/36 + 21/36 - 9/36 = 30/36$

8. b. $P(S) + P(C) + P(PC) = .2 + .25 + .3$

9. d. $P(\text{2 Heads}) \times P(\text{2 Tails}) = (.25)(.25)$

10. $P(A \cap B) = P(A) \cdot P(B|A) = (1/12)(1/11)$

11. positive

12. 1; sample space

13. ≤ 1

14. mutually exclusive

15. a. dependent
    b. independent
    c. independent

16. joint event

17. joint probability

18. conditional probability

19. mutually exclusive; one

20. long-run theoretical probability distribution

21. The likelihood of any event occurring is determined by chance.

22. Any events that are not causally related.

24. A joint probability table is set up in matrix form with rows and columns. For each event in a sample space, the probability is shown for the intersection of the row and column events. For any given row or column, the joint probabilities must sum to the marginal probabilities.

SAMPLE EXERCISE

An agricultural test station was experimenting with the use of herbicides. In one experiment some test plots were treated with herbicide; others were not. A plot was categorized as successful if 75% of the weeds died prior to going to seed.

Joint Probability Table: Success in Killing Weeds with Herbicide

	Treated (T)	Untreated (T')	Total
Success (S)	.30	.35	.65
Failure (S')	.10	.25	.35
Total	.40	.60	1.00

1. The marginal probabilities show that 60% of the test plots were treated with herbicide. Overall, 65% of the plots were categorized as successful, 35% as unsuccessful.

2. The question of interest is: Did the use of the herbicide have a significant impact on weed growth? In other words, were the events *dependent*?

From the table we know that

$P(S \cap T) = .30$
$P(T) = .40$

From Eq. (5-7)

$$P(S|T) = \frac{P(S \cap T)}{P(T)}$$

The probability that a test plot would be categorized a success, given that it had been treated with herbicide, is equal to the probability that the plot was treated *and* categorized as successful divided by the probability that the plot was treated.

$$P(S|T) = \frac{.30}{.40} = .75$$

3. How does this compare with the untreated plots?

From the table we know that

$P(S \cap T') = .35$
$P(T') = .60$

$$P(S|T') = \frac{P(S \cap T')}{P(T')} = \frac{.35}{.60} = .5833$$

4. It appears that the use of the herbicide was related to the categorization of a test plot as (S) or (S'). Further consideration of the dependence or independence of events will be presented in Section 12.4.

CASE PROBLEM

Mr. Garvey, the manager of Garvey Hardware, has been considering changing methods of displaying certain types of hardware, such as nuts, bolts, and nails. In the past the store has used bins containing loose pieces. Lately he has considered switching to the use of small plastic bags and similar packages for the hardware. To check on the impact on sales, he decided to experiment by switching a part of three types of hardware and comparing the results with past experience. Sales that were above past sales were rated good (G) and assigned a numerical value of 3. Sales that were roughly equal to past sales were rated fair (F) and assigned a value of 2. Sales below past sales were rated poor (P) and assigned a value of 1.

After one month Mr. Garvey came up with the following results.

Hardware Type A

Sales Results	Bags (B)	Bins (B')	Total
Good (G)	.17	.18	.35
Fair (F)	.10	.15	.25
Poor (P)	.13	.27	.40
Totals	.40	.60	1.00

Hardware Type B

Sales Results	Bags (B)	Bins (B')	Total
Good (G)	.20	.25	.45
Fair (F)	.15	.20	.35
Poor (P)	.10	.10	.20
Totals	.45	.55	1.00

Hardware Type C

Sales Results	Bags (B)	Bins (B')	Total
Good (G)	.18	.30	.48
Fair (F)	.10	.10	.20
Poor (P)	.07	.25	.32
Total	.35	.65	1.00

Using the statistical concepts learned up to this point, how would you go about deciding if bags or bins were preferred by customers? Additional concepts related to this type of problem will be discussed in Chapter 14.

6

EXPECTED VALUE AND POPULATION PARAMETERS

CHAPTER LEARNING OBJECTIVES

Upon completion of this chapter the student should be able to

- Define the concept of expected value.
- Define the terms discrete and continuous probability distributions.
- Compute expected value for a discrete random variable.
- Define and calculate the population mean and variance.

SYNOPSIS

Expected Value

In earlier chapters a formula for the mean was given as

$$\bar{X} = \frac{\sum_{j=1}^{m} x_j FR(x_j)}{n}$$

where x_j is a midpoint, $FR(x_j)$ is the frequency of x_j occurring, and n is the total number of items. The first term of the summation can be expressed as $x_1 FR(x_1)/n$. Since $FR(x_1)$ represents the number of times that x_1 occurs and n is the total number of outcomes, $FR(x_1)/n$ is, by definition, the relative frequency of x_1. Accordingly, we could replace $FR(x_1)/n$ by $RF(x_1)$. Doing so results in the expression $x_1 RF(x_1)$. Making this equivalent replacement for all terms of the summation, we find that

$$\bar{X} = \sum_{j=1}^{m} x_j RF(x_j).$$

The formula for the expected value of X is similarly expressed:

$$E(X) = \sum_{j=1}^{m} x_j P(x_j)$$

where $P(x_j)$ is the *probability* that x_j will occur. Thus expected value is nothing more than a way to compute an average. This form of the formula is particularly useful in that it can be easily applied to a probability distribution, whereas the $\bar{X}$ formula could not. Recall that a probability distribution can be regarded as a list of all possible unique outcomes and their associated probabilities. To calculate the mean of a probability distribution, we simply multiply the outcome (a random variable) by its associated probability for each possibility and add up the values.

Population Mean and Variance

The mean or expected value has a slightly different interpretation when applied to a probability distribution. The interpretation is that if we repeated the experiment a number of times and took an average, we would expect that average to be close to our calculated expected value. In fact, if we took repeated samples, we would expect the mean values to cluster around the expected values. If we could somehow take an infinite number of trials, the average would be equal to the expected value.

For instance, if asked how many heads would occur on 100 flips of a coin, you would probably respond 50 (this happens to be the expected value). You would also probably recognize that if you actually performed the experiment, the chance of getting exactly 50 heads would be small, whereas the chance of getting close to 50 heads would be high. The most likely result is 50 heads; and if we could repeat the experiment an infinite number of times, our average would be 50.

Note that the expected value implies that there is a true mean. If, somehow, we could measure all items in the population, then the mean would be the expected value. Each sample is simply an estimate of this true mean. The true mean is, in fact, the same as the population mean. This same concept applies to the variance, since the variance is also an average (the average squared error). The expected value of $(X - \mu)^2$ is the population variance. Mathematical manipulation leaves us with the formula

$$E(X - \mu)^2 = \Sigma X^2 P(x) - \mu^2.$$

Probability distributions can be classified as discrete or continuous. A discrete distribution can be thought of as a distribution in which we can count the number of outcomes. All probabilities that have been dealt with so far in this book have been calculated by counting and have been discrete probabilities. A characteristic of discrete distributions is that the random variable is invariably an integer or a categorization. We speak of success as the number of successes, the number on a die, or the draw of a heart.

A continuous distribution, conversely, is one in which we cannot count the number of outcomes. The key to recognizing that we have a continuous distribution is that the random variable can take on values other than integers. Continuous variables involve such measurement as height, weight, temperature, and length. The number of outcomes cannot be counted in that, theoretically, an infinite number of real numbers are possible. Consider the task of counting the number of outcomes for the weight of an adult person. We might well agree that there are limits on the possible value to the weight. Certainly weight can't be less than zero. Perhaps as a maximum we might agree on 500 pounds. If we measure in pounds, then there are 500 possible outcomes. However, we all recognize that we could measure in tenths of pounds, which gives us 5000 outcomes, or hundredths, which gives 50,000 outcomes. If we ignore the physical limitations of weighing devices, we can continue the process of going from tenths to hundredths to thousandths forever. In this sense, an infinite number of outcomes are possible.

Of course, it is recognized that we have ignored a real-world limitation in that weighing devices have some limit. This is true and illustrates one fact: all continuous distributions are theoretical distributions. They are a mathematical formula that is useful in describing reality.

The concept of probability with discrete distributions is the familiar one of counting successes and dividing by the total number of outcomes. With continuous distributions, this concept has little meaning, for there are an infinite number of outcomes; and whenever we divide by infinity, we end up with a probability of zero. Any point probability for a continuous distribution is zero. As a result, our interest in continuous probability distributions is to consider the probability of intervals. How to calculate it will be left to later chapters.

KEY TERMS

Continuous probability distribution	Distribution of a variable that may assume any value of a continuous scale.
Discrete probability distribution	Distribution of a variable that may only assume a finite or countable number of numerical values.
Expected value	Weighted average of the variable's possible values, using the prospective probabilities as weights.
Fair game	Game of chance in which no player is favored.
Finite population	Population containing a fixed number of units or observations.
Infinite population	Population that does not have a fixed number of units or observations, or in which N approaches infinity.
Parent population	Population of the original random variable X that generates other probability distributions.
Sample distribution	Frequency distribution of a sample.

Variance   The average of the squared deviations from the mean.

FORMULA REVIEW

(6-1) Mathematical expectation

$$E(X) = \sum_{j=1}^{m} x_j P(x_j) = x_1 P(x_1) + x_2 P(x_2) + \cdots x_m P(x_m)$$

(6-2) Population mean

$$\mu = \frac{\sum_{i=1}^{N} X_i}{N}$$

(6-4) Population variance

$$\sigma^2 = E(X^2) - \mu^2 = \Sigma\, x^2 P(x) - \mu^2$$

ORGANIZED LEARNING QUIZ

1. What is a fair price to pay to enter a game in which you can win $20 with a probability of .3?

   a. $1.00
   b. $3.50
   c. $6.00
   d. $7.50

2. If the wind blows, a wind-damage insurance salesman can earn $100 per day. If the weather is calm, he can lose $10 per day. What is his expected value if the probability of wind is .25?

   a. $22.50
   b. $17.50
   c. $15.50
   d. $12.50

3. The local fire department is selling 1000 chances, at $2 each, to win an $800 trail bike. This is a fair game.

   a. True
   b. False

4. Let X be the number of heads obtained in the toss of five coins. What is the expected value of X? (Let H = 1 and T = 0.)

   a. 2.0
   b. 2.25
   c. 2.50
   d. 3.0

5. Mr. Barry is considering investing in gold futures. He has $10,000 to invest. Let X designate the amount of money he will end up with; it has the following probability distribution.

X	P(x)
$10,000	.2
15,000	.3
20,000	.2
7,500	.3

What can Mr. Barry expect if he decides to invest $10,000?

a. $10,000
b. $12,000
c. $12,750
d. $13,500

6. Suppose that a bowl contains ten pieces of fruit, two of which are inedible. Two pieces are removed for consumption. Let X be the number of bad pieces of fruit in the sample of two. What is the expected value of X?

a. .64
b. .36
c. .40
d. .60

7. Mr. Kirkwood is considering investing in the development of a new carburetor. Let X be the amount of money he will earn and P(x) be the true probability that the carburetor will be a success.

X	P(x)
$100,000	.1
110,000	.2
120,000	.3
50,000	.4

What is the most that Mr. Kirkwood should be willing to invest to ensure that he at least breaks even?

a. $88,000
b. $50,000
c. $70,500
d. $85,000

8. Following is the probability distribution for the number of daily calls received by a police dispatch unit.

Phone calls (x)	Probability
10	.1
11	.2
12	.2
13	.2
14	.2
15	.1

What is the expected daily number of calls received?

    a. 11.2
    b. 13.2
    c. 12.0
    d. 12.5

9. What is the variance of X in problem 8?

    a. 2.00
    b. 2.25
    c. 2.50
    d. 2.8

10. A sheep rancher raises two different breeds of sheep. The following probability distributions exist for the likelihood that a ewe will have a given number of lambs.

Breed A		Breed B	
x	P(x)	x	P(x)
0	.2	0	.1
1	.3	1	.4
2	.4	2	.5
3	.1		

What are the expected number of births for the two breeds?

    a. 1.6, 1.4
    b. 1.5, 1.6
    c. 1.4, 1.4
    d. 1.5, 1.5

11. A game is fair when no one player is _____.

12. The expected value of participating in a fair game is _____.

13. The expected value of a discrete random variable is the weighted average of the values that the variable may assume over the _____.

14. It is possible to calculate the population mean of an infinite population if values can be grouped into a _____.

15. An infinite population can be generated from a finite set of values or units if sampling is taken with _____.

16. The population standard deviation is a measure of the variation of all values of random variable (X) from the _____.

17. The _____ is the average of the squared deviations from the mean.

18. The population from which other probability distributions is

generated is called the _____.

19. _____ deals with the methods of making inferences about population parameters based on _____ statistics.

20. The distribution of values for a sample statistic, such as the sample mean, is called a _____.

21. Define expected value.

_____

_____

_____

22. Define the term population parameters.

_____

_____

_____

23. Describe the two major population parameters.

_____

_____

_____

ANSWERS TO ORGANIZED LEARNING QUIZ

1. c. $E(X) = .3(20)$

2. b. $E(X) = .25(100) - .75(10)$

3. b. $E(X) = 798(\frac{1}{1000}) - 2(\frac{900}{1000}) = -1.20$

4. c. $E(X) = 0P(0) + 1P(1) + 2P(2) \cdots$

5. c. $E(X) = .2(10,000) + .3(15,000) + .2(20,000) + .3(7500)$

6. c. $E(X) = (\frac{8}{10})(\frac{7}{9})(0) + (\frac{8}{10} \cdot \frac{2}{9} + \frac{2}{10} \cdot \frac{8}{9})(1) + (\frac{2}{10})(\frac{1}{9})(2)$

7. a. $E(X) = .1(100,000) + .2(110,000) + .3(120,000) + .4(50,000)$

8. d. $E(X) = .1(10) + .2(11) + .2(12) + .2(13) + .2(14) + .15(.1)$

9. b. $\sigma^2 = \Sigma x^2 P(x) - \mu^2 = 158.5 - (12.5)^2 = 158.5 - 156.25$

10. c.

11. favored

12. zero

13. long run

14. distribution

15. replacement

16. population mean (μ)

17. variance

18. parent population

19. statistical inference, sample

20. sampling distribution

21. Expected value of a discrete random variable is the sum of the value of each outcome multiplied by the probability of its occurrence.

22. Population parameters are summary measures of particular population properties.

23. The two most frequently used population parameters are the mean and variance. As with the sample mean, the population mean μ is a measure of central tendency; the variance is a measure of dispersion.

SAMPLE EXERCISE

Examine the following table and calculate the expected value and variance for the random variable X.

x	x^2	P(x)	$x^2 P(x)$
10	100	.10	10
20	400	.20	80
30	900	.30	270
40	1600	.40	640
		1.00	1000

1. Calculate the mean.

$$E(X) = \sum_{j=1}^{m} x_j P(x_j) = x_1 P(x_1) + x_2 P(x_2) + \cdots + x_m P(x_m)$$

$$E(X) = 10(.10) + 20(.20) + 30(.30) + 40(.40)$$

$$= 1 + 4 + 9 + 16$$

$$= 30$$

2. Calculate the variance.

$$\sigma^2 = E(X^2) - \mu^2 = \Sigma x^2 P(X) - \mu^2$$

$$\sigma^2 = 1000 - 30^2 = 1000 - 900 = 100$$

$$\sigma = \sqrt{\sigma^2} = 10, \text{ the standard deviation}$$

CASE PROBLEM

The church bazaar has designed a game to raise money. The game consists of an individual paying the game fee and selecting a colored ball from a vase. The payoff depends on the color of ball drawn. The table below shows the distribution of the balls and the payoff associated with each. The ball is replaced after each play so that there are always six balls in the vase at the time an individual draws.

Color	Number of Balls	Payoff (X)
Red	3	.25
Blue	2	.50
White	1	1.00

Find the expected value of the game and the variance of the variable X.

7

DISCRETE PROBABILITY DISTRIBUTIONS

CHAPTER LEARNING OBJECTIVES

Upon completion of this chapter the student should be able to

- Distinguish between binomial, hypergeometric, and Poisson probability distributions.

- Interpret the nature of the binomial, hypergeometric, and Poisson probability functions.

- Interpret the meaning of the mean and variance of the Bernoulli random variable and binomial random variable.

- Interpret the meaning of the success/failure terms in describing the outcomes of the binomial and Poisson distributions.

- Calculate the probability of a given outcome occurring in n trials for each distribution.

SYNOPSIS

A discrete random variable is a variable that assumes a finite set of values that typically are nonnegative. Here a number of useful discrete probability distributions--the binomial, hypergeometric, and Poisson distributions--are covered. By way of introduction, the Bernoulli process, which forms the basis of the binomial distribution, is explained.

Bernoulli Model

A Bernoulli process is a series of experiments or events, each of which may take on one of two possible outcomes. These outcomes are defined somewhat arbitrarily as "success" and "failure." Because they are qualitative in nature, the outcomes are assigned values of 1 (success) and 0 (failure). If p equals the probability of success, $1 - p = q$ = the probability of a failure. The expected value E(W) for a Bernoulli random variable is p. The variance

σ_w^2 is pq.

Binomial Random Variable

As stated, the Bernoulli model forms the basis for the binomial random variable. The random variable X is the number of successes in n trials taken by a Bernoulli process. Since X is defined as the number of successes in n trials, each with an expected value p, the expected value or mean of the random variable X is np, the number of trials times the expected value of one trial. The variance of the sum of a set of independent variables W is equal to the sum of the variance of the variables. From this is derived the variance of X: npq.

The probability distribution of a variable is the set of possible outcomes and the probability associated with each outcome. Since the number of possible outcomes becomes very large as n increases, listing all outcomes would be tedious. The rule of combination can be applied to calculate the number of outcomes. The coefficient, or number of simple random events for each value, for the event of x successes in n trials is C_x^n. Using this, we can derive the probability that any given event will occur.

Hypergeometric Distribution

The binomial distribution is based on the assumption that the possible observations are independent of one another. Where this assumption does not hold, a different distribution model is used-- the hypergeometric distribution. This model is used when the probability changes with each successive observation. Selecting a king from a normal deck of cards changes the probability that a king will be selected on the next trial. Not only has the number of successes (kings) remaining in the deck changed but the size of the population has also gone down by one.

Poisson Distribution

The third distribution--the Poisson-- is applicable to many processes that generate observations per unit of time or space. This is equivalent to the binomial "number of successes in n trials." The Poisson distribution has mean μ = np. The standard deviation is $\sqrt{np}$ or $\sqrt{\mu}$. The distribution has only the one parameter, μ. There are a large number of applications for the Poisson distribution in business, where the concern is with the occurrence of certain events per unit of time or space.

KEY TERMS

Bernoulli model	Probability model of a Bernoulli random variable.
Bernoulli trials	An experiment that can lead to only one of two possible outcomes.
Binomial coefficients	The number of simple random events for each value of the binomial random variable.

Binomial distribution	Probability distribution that has two parameters n and p, where n, the number of trials, must equal two or more.
Binomial random variable	Number of successes in n trials taken by a Bernoulli process.
Cumulative probability distribution	Function showing the probability equal to or smaller than each of the possible values.
Hypergeometric distribution	Sampling from a finite population without replacement so that the observations are not independent and the probability of success changes for each successive observation.
Mean of X	np; probability of success times the number of trials.
Poisson exponential distribution	Samples drawn from a process that generates observations per unit of time or space.
Population proportion	Probability of a success, or the proportion of successes obtained in the long run.
Sample proportion	Probability of a success within a sample; proportion of successes in n trials.
Success, failure	Terms used to identify outcomes of a Bernoulli trial; they do not indicate anything about the relative desirability of the outcomes.
Two-point model	Probability model having only two classes of events.
Variance of X	The variance of the sum of a set of independent variables is equal to the sum of the variances of the variables.

FORMULA REVIEW

(7-1) Mean Bernoulli random variable $\quad E(W) = p$

(7-2) Variance Bernoulli random variable $\quad \sigma_W^2 = pq$

(7-3) Binomial random variable $\quad X = W_1 + W_2 + W_3 + \cdots + W_n = \Sigma W$

(7-4) Mean Binomial random variable $\quad E(X) = np$

(7-5) Variance
Binomial random
variable
$$\sigma_X^2 = npq$$

(7-6) Binomial
probability
function
$$P(X=x) = C_x^n (p)^x (1-p)^{n-x} \text{ for } x = 0, 1, 2, \ldots, \text{ or } n$$

(7-7) Cumulative
probability
function
$$P(X \leq k \mid n, p) = \sum_{x=0}^{k} C_x^n p^x q^{n-x}$$

(7-8) Hypergeometric
probability
distribution
$$P(x) = \frac{C_x^{n_1} C_{n-x}^{n_2}}{C_n^{n_1+n_2}} \text{ for } x = 0, 1, 2, \ldots, n$$

(7-9) Poisson
exponential
distribution
$$P(x) = e^{-\mu} \frac{\mu^x}{x!}$$

ORGANIZED LEARNING QUIZ

Five dice are rolled simultaneously. The outcomes of the rolls are independent. For each die, a success is defined as showing an even number of spots; a failure is defined as landing with an odd number of spots showing.

1. This represents a

   a. binomial random variable.
   b. hypergeometric random variable.
   c. Poisson process.

2. The mean of X, the number of successes obtained in the roll of five dice, is

   a. 1.5.
   b. 2.0.
   c. 2.5.
   d. 3.0.

3. The variance of X is

   a. .75.
   b. 1.0.
   c. 1.25.
   d. 1.5.

4. The probability of no successes P(X=0) is

   a. .125.
   b. .031.
   c. .005.

5. The probability of five successes P(X=5) is

   a. the same as P(X=0).
   b. greater than P(X=0).
   c. Less than P(X=0).

6. Each of the outcomes is a

   a. combination.
   b. permutation.

7. The probability that two or fewer dice will land showing an even number of dots is

   a. .03125.
   b. .15625.
   c. .3125.
   d. the sum of a, b, and c, or .5000.

8. The probability that at least three dice will land showing an even number of dots is

   a. P(X=0) + P(X=1) + P(X=2) + P(X=3).
   b. .5000.
   c. 1 - (.4687) = .5313.

9. If only one die is rolled, the outcome can be defined in such a way that the experiment is a Bernoulli trial.

   a. True
   b. False

10. Suppose that 75% of the students in a class receive a grade of C. Let W be the random variable that takes on a 1 when a randomly selected student is one of those receiving a C and 0 when he or she is not. The mean of W is _____ and the standard deviation is _____.

11. Assume that 20% of all drivers who run a certain red light are drunk. What is the probability that in a sample of 20 drivers who ran the red light 5 are drunk? _____

12. Find the values for the following expressions.

    a. $C_2^4 (.4)^2 (.6)^2$ _____
    b. $C_3^4 (.4)^3 (.6)^1$ _____
    c. $C_0^4 (.4)^0 (.6)^4$ _____

13. The probability of a success in a Bernoulli random variable is a population _____, or the _____ of successes obtained in the long run.

14. The binomial random variable X can be defined as the number of successes on n observations taken by a _____ process.

15. The _____ is the number of simple random events for each value of the binomial random variable.

16. A hypergeometric distribution differs from a binomial distribution in that there is sampling _____ replacement and consequently the _____ changes for each successive observation.

17. A class is made up of ten males and five females. What is the probability that in a committee of three selected at random without replacement both sexes are represented proportionally?

   a. $C^{10}_2 C^5_1 / C^{15}_3$ _____

   b. $C^{10}_3 C^5_3 / C^{15}_5$ _____

   c. $C^{10}_2 / C^{15}_3 + C^5_1 / C^{15}_3$ _____

18. A Poisson distribution is characterized by the number of successes per _____.

19. The Poisson distribution is one for which n is _____ and p is _____. When np increases, the Poisson distribution will be closer and closer to a _____ curve.

20. The Poisson distribution has a μ = _____ and the standard deviation is _____.

21. Give examples of each of the following (other than examples given in the text): binomial, hypergeometric, and Poisson random variables.

   _____
   _____
   _____

22. Explain what the variance of a binomial random variable tells about the distribution.

   _____
   _____
   _____

23. Explain the value of the cumulative probability function.

   _____
   _____
   _____

24. Discuss the meaning of the outcomes of a Poisson distribution.

_____

_____

_____

25. Discuss the nature of the outcomes of a hypergeometric distribution.

_____

_____

_____

ANSWERS TO ORGANIZED LEARNING QUIZ

1.	a	2.	c
3.	c	4.	b
5.	a	6.	a
7.	d	8.	b
9.	a	10.	.75, .43
11.	.0873	12.	a. .3456; b. .1536; c. .1296
13.	proportion, proportion	14.	Bernoulli
15.	binomial coefficient	16.	without, probability
17.	a	18.	unit of time or space
19.	large, small, bell-shaped	20.	$np, \sqrt{\mu}$

21. Binomial: herbicide kills (success) or fails to kill (failure) an adequate proportion of a certain type of plant; hypergeometric: many types of production sampling, or sampling where population proportions are known; Poisson: any process that generates observations per unit of time or space.

22. As with other distributions, the variance is an expected value of a function. It measures the deviations from the expected value of the variable.

23. Enables us to calculate the probabilities in the forms of less than or equal to, more than or equal to, at least, at most, and so on.

24. The Poisson distribution is a probability distribution of outcomes that are measured per unit of time or space, as in shoppers arriving per minute.

25. The hypergeometric distribution is a probability distribution of outcomes taken from a finite population without replacement. The observations are related and the probabilities change after each observation.

SAMPLE EXERCISE

Consult the Sample Exercise at the end of Chapter 3 of this study guide (p. 27 - 29). Assume that the management of Company L is interested in taking a preliminary look at salaries by comparing the salaries of each of their 13 managers with the average (mean) salary of Company G. They have decided to define a success as a salary that exceeds the G company mean.

1. What is the proportion of successes for Company L?

2. Assume that the true (population) proportion equals .5. What is the probability that, of a sample of four managers, two will have salaries above the G company mean?

ANSWERS TO SAMPLE EXERCISE

1. $\text{Mean}_G = 510$, $p = .46$

2. $P(X = 2 | n = 4, p = .5) = C_2^4 (.5)^2 (.5)^{4-2} = \dfrac{4!}{(4-2)!2!} (.5)^4$

$= 6(.0625) = .375$

CASE PROBLEM

Smith City Airport is a medium-sized facility in a growing city. Airline representatives have become increasingly concerned that existing baggage handling equipment is inadequate.

Assume that 10 minutes or less is considered adequate (10 minutes or less is a success).

1. If the true proportion of baggage delays is .4, what is the probability that, in a sample of ten, four will be delayed?

2. The administrator collected the following data from the flights for a particular day.

Flight number	Approximate Elapsed Time from Unloading of Luggage to Arrival of Luggage in Customer Pick-up Area (minutes)
132	9
138	15
412	28
517	13
215	8
220	12
415	9
310	15
135	12
216	10

Find $E(X)$ and σ_X^2

3. What is the probability that on any given day only three flights will experience baggage delays (i.e., greater than 10 minutes)? Five flights? At least five flights?

ANSWERS TO CASE PROBLEM

1. $P(X = 4) = \dfrac{10!}{4!(10-4)!} (.4)^4 (1-.4)^{10-4}$

   $= 210(.4)^4(.6)^6$

   $= 210(.0256)(.046656)$

   $= .2508$

2. $E(X) = np = 10(.4) = 4$

   $\sigma_{X^2} = npq = 10(.4)(.6) = 2.4$

3. $P(X = 3) = .215$

   $P(X = 5) = .201$

   $P(X = \text{at least } 5) = P(X>4) = 1 - P(X\leq 4) = 1 - .6631 = .3369$

COMPUTER PROGRAMS

    The BASIC computer program provided computes a binomial, hypogeometric (two groups), or Poisson probability. It can be a point probability or one of three cumulative probability options. The required inputs are: selection of the probability distribution, probability type, probability of success, sample size, the population size for the hypogeometric probability distribution, and the number of successes (R) for which a probability is to be calculated. All input is prompted by the program.

```
10 REM************************************************************
20 REM
30 REM      TITLE:      PROBCALC
40 REM
50 REM      LANGAUGE:   BASIC
60 REM
70 REM      DESCRIPTION:
80 REM         THIS PROGRAM CONTAINS FOUR DIFFERENT
90 REM         STANDARD PROBABILITY DISTRIBUTION
100 REM        CALCULATIONS.  IT CAN BE A POINT
110 REM        PROBABILITY OR ONE OF THREE CUM-
120 REM        ULATIVE PROBABILITY OPTIONS.
130 REM
140 REM
150 REM     INSTRUCTIONS:
160 REM        ENTER DATA AS REQUESTED BY THE PROGRAM.
170 REM
180 REM************************************************************
190 REM
200 KEY OFF
210 PRINT:PRINT:PRINT:PRINT:PRINT:PRINT:PRINT:PRINT:PRINT:PRINT
220 PRINT"             PROGRAM TO COMPUTE PROBABILITIES"
230 PRINT:PRINT
240 PRINT "                     MENU"
250 PRINT
260 PRINT "         1.  BINOMIAL DISTRIBUTION":PRINT
270 PRINT "         2.  HYPOGEOMETRIC DISTRIBUTION":PRINT
280 PRINT "         3.  POISSON DISTRIBUTION":PRINT
290 PRINT "         4.  HYPOGEOMETRIC DISTRIBUTION"
300 PRINT "              MULTIGROUPS":PRINT
310 PRINT "         0.  EXIT PROGRAM":PRINT:PRINT:PRINT:PRINT
320 PRINT "WHAT IS YOUR CHOICE?   (TYPE 1,2,3,4, OR 0 AND RETURN)";
```

```
330 INPUT K1
340 CLS
350 IF K1=0 THEN 2280
360 IF K1=4 THEN 1600
370 IF K1=2 THEN 430
380 IF K1=3 THEN 450
390 IF K1=1 THEN 410
400 GOTO 320
410 PRINT:PRINT:PRINT"PROGRAM TO COMPUTE BINOMIAL PROBABILITY"
420 GOTO 470
430 PRINT:PRINT:PRINT "PROGRAM TO COMPUTE HYPOGEMETRIC PROBABILITY"
440 GOTO 470
450 PRINT:PRINT:
460 PRINT "PROGRAM TO COMPUTE POISSON PROBABILITY"
470 PRINT:PRINT "PROBLEM NUMBER";
480 INPUT K9
490 PRINT "SELECT OPTION"
500 PRINT "   1: P(X=R)         2: P(X <= R)"
510 PRINT "   3: P(X >= R)      4: P(R1 <= X <= R2)";
520 INPUT K2
530 IF K2<1 OR K2 > 4 THEN 490
540 PRINT "PROBABILITY OF SUCCESS";
550 INPUT P1
560 IF P1 <= 0 OR P1 >= 1 THEN 540
570 N1=1000000!
580 IF K1<>2 THEN 660
590 PRINT "POPULATION SIZE";
600 INPUT N1
610 IF INT(P1*N1)=P1*N1 THEN 660
620 PRINT "ERROR--FOR A HYPOGEOMETRIC PROBABILITY DISTRIBUTION"
630 PRINT "THE PROBABILITY OF SUCCESS TIMES THE POPULATION SIZE"
640 PRINT "MUST EQUAL AN INTEGER.  Please Input Your Data Again"
650 GOTO 540
660 PRINT "SAMPLE SIZE";
670 INPUT N
680 IF N<=0 OR N>=N1 THEN 660
690 IF K2=4 THEN 740
700 PRINT "R=";
710 INPUT R
720 IF R<0 OR R>N THEN 700
730 GOTO 800
740 PRINT "R1=";
750 INPUT R1
760 IF R1<0 OR R1>N THEN 740
770 PRINT "R2=";
780 INPUT R2
790 IF R2<=R1 OR R2>N THEN 770
800 IF K2<>1 THEN 850
810 R3=R
820 GOSUB 1510
830 P2=P
840 GOTO 1010
850 IF K2<>2 THEN 890
860 A=1
870 B=R+1
880 GOTO 950
890 IF K2<>3 THEN 930
900 A=R+1
910 B=N+1
920 GOTO 950
930 A=R1+1
940 B=R2+1
950 P2=0
960 FOR I1=A TO B
970 R3=I1-1
980 GOSUB 1510
990 P2=P2+P
1000 NEXT I1
1010 IF K2<>1 THEN 1040
1020 PRINT "P(X=";R;")=";P2
1030 GOTO 1110
1040 IF K2<>2 THEN 1070
1050 PRINT "P(X<=";R;")=";P2
1060 GOTO 1110
1070 IF K2<>3 THEN 1100
1080 PRINT "P(X>=";R;")=";P2
```

```
1090 GOTO 1110
1100 PRINT "P(";R1;"<=X<=";R2;")=";P2
1110 PRINT
1120 PRINT
1130 LOCATE 24,1:PRINT "TO CONTINUE PRESS ANY KEY":A$=INPUT$(1)
1140 GOTO 200
1150 REM SUBROUTINE TO COMPUTE BINOMIAL PROBABILITY
1160 GOSUB 1190
1170 P=P*(P1^R3)*((1-P1)^(N-R3))
1180 RETURN
1190 REM SUBROUTINE TO EVALUATE R3 FROM N COMBINATIONS
1200 P=1
1210 IF R3=0 THEN 1250
1220 FOR I=1 TO R3
1230 P=P*(N-I+1)/I
1240 NEXT I
1250 RETURN
1260 REM SUBROUTINE TO CALCULATE HYPOGEOMETRIC PROBABILITY
1270 N4=N
1280 R4=R3
1290 N=INT(P1*N1+.5)
1300 P3=1
1310 GOSUB 1190
1320 P3=P3*P
1330 R3=N4-R3
1340 N=N1-N
1350 GOSUB 1190
1360 P3=P3*P
1370 R3=N4
1380 N=N1
1390 GOSUB 1190
1400 P=P3/P
1410 RETURN
1420 REM SUBROUTINE TO CALCULATE POISSON PROBABILITY
1430 M=N*P1
1440 P=1
1450 IF R3=0 THEN 1490
1460 FOR I=1 TO R3
1470 P=P*I
1480 NEXT I
1490 P=((2.7182^(-M))*(M^R3))/P
1500 RETURN
1510 REM SELECT SUBROUTINE TO CALCULATE PROBABILITY
1520 IF K1<>1 THEN 1550
1530 GOSUB 1150
1540 GOTO 1590
1550 IF K1<>2 THEN 1580
1560 GOSUB 1260
1570 GOTO 1590
1580 GOSUB 1420
1590 RETURN
1600 PRINT:PRINT:
1610 PRINT "PROGRAM--HYPOGEOMETRIC PROBABILITY--MULTI GROUPS"
1620 PRINT
1630 PRINT
1640 PRINT"                       N(1)    N(2)              N(N)"
1650 PRINT"                      (   )  (   ) ........(    )"
1660 PRINT"                       R(1)    R(2)              R(N)"
1670 PRINT"P(R(1),R(2),.,R(N))=---------------------------"
1680 PRINT"                       (N(1)+N(2)+......+N(N)"
1690 PRINT"                      (                      )"
1700 PRINT"                       (R(1)+R(2)+......+R(N)"
1710 PRINT
1720 DIM N(15),X(15),S(15),M(15),C(15)
1730 PRINT
1740 PRINT
1750 PRINT"PROBLEM NUMBER"," ",
1760 INPUT N2
1770 PRINT"NUMBER OF GROUPS"," ",
1780 INPUT N2
1790 IF N2<=0 OR N2>15 THEN 1750
1800 FOR J=1 TO N2
1810 PRINT"         TOTAL NUMBER IN GROUP";J,
1820 INPUT N(J)
1830 IF N(J)<0 THEN 1810
1840 PRINT"NUMBER OF SUCCESSES IN GROUP";J,
1850 INPUT X(J)
1860 IF X(J)<0 OR X(J)>N(J) THEN 1840
```

```
1870 NEXT J
1880 PRINT
1890 PRINT
1900 GOSUB 2050
1910 PRINT
1920 PRINT
1930 PRINT"CONTINUE? 1=YES, 0=NO";
1940 INPUT I
1950 IF I=1 THEN 1730
1960 LOCATE 24,1:PRINT "TO CONTINUE PRESS ANY KEY";:A$=INPUT$(1)
1970 GOTO 200
1980 REM--SUBROUTINE--EVALUATE R3 FROM N COMBINATIONS
1990 P=1
2000 IF R3=0 THEN 2040
2010 FOR I=1 TO R3
2020 P=P*(N-I+1)/I
2030 NEXT I
2040 RETURN
2050 REM--SUBROUTINE--CALCULATE HYPOGEOMETRIC PROBABILITY
2060 S1=0
2070 S2=0
2080 P3=1
2090 PRINT "P=";
2100 FOR J=1 TO N2
2110 N=N(J)
2120 R3=X(J)
2130 S1=S1+R3
2140 S2=S2+N
2150 GOSUB 1980
2160 IF J=N2 THEN 2190
2170 PRINT P;"X";
2180 GOTO 2200
2190 PRINT P;"/";
2200 P3=P3*P
2210 NEXT J
2220 N=S2
2230 R3=S1
2240 GOSUB 1980
2250 PRINT P;"=";P3/P
2260 P=P3/P
2270 RETURN
2280 KEY ON:END
```

```
            PROGRAM TO COMPUTE PROBABILITIES

                        MENU

            1.   BINOMIAL DISTRIBUTION

            2.   HYPOGEOMETRIC DISTRIBUTION

            3.   POISSON DISTRIBUTION

            4.   HYPOGEOMETRIC DISTRIBUTION
                    MULTIGROUPS

            0.   EXIT PROGRAM


WHAT IS YOUR CHOICE?   (TYPE 1,2,3,4, OR 0 AND RETURN)? 1


PROGRAM TO COMPUTE BINOMIAL PROBABILITY

PROBLEM NUMBER? 1
SELECT OPTION
   1: P(X=R)             2: P(X <= R)
   3: P(X >= R)          4: P(R1 <= X <= R2)? 2
PROBABILITY OF SUCCESS?  .5
SAMPLE SIZE? 10
R=? 5
P(X<= 5 )= .6230469
```

```
TO CONTINUE PRESS ANY KEY

            PROGRAM TO COMPUTE PROBABILITIES

                         MENU

            1.   BINOMIAL DISTRIBUTION

            2.   HYPOGEOMETRIC DISTRIBUTION

            3.   POISSON DISTRIBUTION

            4.   HYPOGEOMETRIC DISTRIBUTION
                     MULTIGROUPS

            0.   EXIT PROGRAM



WHAT IS YOUR CHOICE?   (TYPE 1,2,3,4, OR 0 AND RETURN)? 2


PROGRAM TO COMPUTE HYPOGEOMETRIC PROBABILITY

PROBLEM NUMBER? 2
SELECT OPTION
  1: P(X=R)           2: P(X <= R)
  3: P(X >= R)        4: P(R1 <= X <= R2)? 1
PROBABILITY OF SUCCESS? .5
POPULATION SIZE? 20
SAMPLE SIZE? 10
R=? 2
P(X= 2 )= .0109604


TO CONTINUE PRESS ANY KEY

            PROGRAM TO COMPUTE PROBABILITIES

                         MENU

            1.   BINOMIAL DISTRIBUTION

            2.   HYPOGEOMETRIC DISTRIBUTION

            3.   POISSON DISTRIBUTION

            4.   HYPOGEOMETRIC DISTRIBUTION
                     MULTIGROUPS

            0.   EXIT PROGRAM



WHAT IS YOUR CHOICE?   (TYPE 1,2,3,4, OR 0 AND RETURN)? 4



PROGRAM--HYPOGEOMETRIC PROBABILITY--MULTI GROUPS

                      N(1)    N(2)              N(N)
                    (    ) (      ) ........(       )
                      R(1)    R(2)              R(N)
P(R(1),R(2),.,R(N))=------------------------------
                         (N(1)+N(2)+......+N(N) )
                         (R(1)+R(2)+......+R(N) )
```

```
PROBLEM NUMBER                              ? 4
NUMBER OF GROUPS                            ? 3
        TOTAL NUMBER IN GROUP 1             ? 5
NUMBER OF SUCCESSES IN GROUP 1              ? 2
        TOTAL NUMBER IN GROUP 2             ? 10
NUMBER OF SUCCESSES IN GROUP 2              ? 4
        TOTAL NUMBER IN GROUP 3             ? 12
NUMBER OF SUCCESSES IN GROUP 3              ? 6


P= 10 X 210 X 924 / 1.738386E+07 = .1116208


CONTINUE? 1=YES, 0=NO?0




            PROGRAM TO COMPUTE PROBABILITIES

                        MENU

            1.   BINOMIAL DISTRIBUTION

            2.   HYPOGEOMETRIC DISTRIBUTION

            3.   POISSON DISTRIBUTION

            4.   HYPOGEOMETRIC DISTRIBUTION
                    MULTIGROUPS

            0.   EXIT PROGRAM



WHAT IS YOUR CHOICE?  (TYPE 1,2,3,4, OR 0 AND RETURN)? 0
OK
```

8

THE NORMAL PROBABILITY DISTRIBUTION

CHAPTER LEARNING OBJECTIVES

Upon completion of this chapter the student should be able to

- Define characteristics of the normal probability distribution.

- Differentiate between a discrete and continuous random variable.

- Calculate a Z score.

- Calculate probabilities from a normal probability distribution, using the normal probability tables.

- Define a sampling distribution.

- Define the central limit theorem.

- Calculate probabilities from the binomial distribution, using the normal approximation.

- Define the conditions necessary for the normal probability distribution to be a good approximation of the binomial probability distribution.

SYNOPSIS

Nature of the Normal Distribution

The normal probability distribution represents the single most important probability distribution in statistics. It is important for a number of reasons. The normal distribution closely represents the distribution of a great many random variables that are commonly used in real life. Secondly, other important probability distributions have been derived that require the assumption of normality for their measurements. Next, the normal distribution can be used as an approximation for other probability distributions under conditions where the original distribution would be difficult to use.

Finally, the central limit theorem suggests that some critical sampling distributions approach normality for sample sizes well within practical limits. This statement is true regardless of the shape of the original distribution.

It is important to recognize that a normal probability distribution is a theoretical distribution. In other words, it is mathematically derived and not necessarily based on physical reality. This, of course, is true of all theoretical distributions. Normal distributions do not, in fact, exist. Almost all things in physical reality that we choose to represent with a normal probability distribution do not fully meet the mathematical requirements of a normal distribution. Nevertheless, many items can be usefully represented by the normal distribution, and it is, in fact, a highly useful distribution.

The physical process that would lead to a normal distribution can be thought of this way. Let us consider a part we are machining. The part has some average width that the machine is set to produce. We know that the machine does not produce this exact width every time, for there are a large number of uncontrolled influences, such as the material, different adjustments on the machine, perhaps air temperature and humidity, and vibration. If these factors are independent of one another and happen basically by chance, and if no one factor is dominant over the other, the resulting distribution of measurements should then be approximately normally distributed. Most of the time, errors will tend to cancel and the measurement will be close to the mean. This is the center part of our bell-shaped curve. Occasionally the errors will be primarily in one direction and result in the observations in the tails of the normal distribution. It's like flipping coins to see the direction of the error. Generally the number of heads and tails will be about even. Occasionally an extreme result occurs.

The normal distribution has certain characteristics. It is a continuous distribution in that the random variable can take on any real number from $-\infty$ to $+\infty$. Any nonzero interval has some positive probability of occurring. The distribution is symmetric with the probability being the highest around the mean and becoming proportionately less as we move toward the tails of the distribution. The median and mode of the distribution are equal to the mean of the distribution. Also, a normal distribution is completely defined by two parameters: the mean and the standard deviation.

The method for calculating probability changes when a continuous distribution rather than a discrete distribution is used. As noted earlier, the point probability of a continuous distribution is zero. Probability is calculated over intervals when we deal with continuous distributions. It is also true that all continuous distributions are theoretical distributions and are defined by a mathematical function (termed a probability density function). The particular function for the normal curve is

$$f(x) = \frac{1}{\sqrt{2\pi\sigma^2}} e^{-(x-\mu)^2/2\sigma^2}.$$

If we were to plot this curve, we would get the familiar bell-shaped curve.

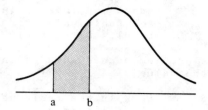

To find the probability of a value falling in the area between a and b, calculate the area under the curve between points a and b (this can be done by integrating the curve from a to b). For convenience, these calculations have been done for you and are tabulated in a table for the special case where $\mu = 0$ and $\sigma = 1$.

The Standard Normal Distribution

For each value of μ and σ, there is a unique normal probability distribution. Making tables for each possibility is impractical. The solution to providing adequate tables is to tabulate one distribution and transform all others to that distribution. The one distribution that we tabulate is called the *standard normal* and has a mean of zero and a standard deviation of one. Any other curve can be converted to the standard normal by calculating a Z score.

$$Z = \frac{X - \mu}{\sigma}$$

Z, in this case, is the equivalent value of X measured in standard deviations from the mean. It follows a standard normal distribution and may be looked up in the standard normal tables.

Some Applications of the Z Score

To calculate a probability involving X, we first calculate its Z value and then look up the Z value in the table and find the appropriate probability. It should be remembered that the total area under the curve is 1 and the area in each half is .5. Also, be sure to remember that the curve is symmetric. Normal probability tables differ in their presentation; so when using a table, be sure to verify which probabilities are in the table. This book provides the area left of Z. Some tables provide the area from Z to the mean for positive Z values only. In each case, it is probable that some manipulation of probabilities will be required. The following guidelines should help for the tables in this book.

$P(X \leq A)$ Calculate the Z value equivalent to A and look up directly in the table.

$P(X \geq A)$ Calculate the Z value equivalent to A and look up in the table. Subtract the table value from 1.

$P(A \leq X \leq B)$ Calculate the Z equivalent values for A and B and look up in the table. Subtract the table value for A from the table value for B.

$P(A \geq X \geq B)$ Calculate the Z equivalent values for A and B and look up in the table. Subtract the table value for B from 1 and add the result to the table value for A.

Another alternative problem occurs when given a probability value and asked to find the value of A.

$P(X \leq A) = .38$ Look in the table value for the probability (in this case, .38) and find its Z value. Then calculate by

$$A = \mu + Z\sigma.$$

$P(X \geq A) = .05$ Subtract the probability (.05) from 1 and look this value up in the table and find its Z value. Then calculate A as above.

$P(A \leq X \leq B) = .90$ For this problem, it must be assumed that A and B are equidistant from the mean. If so, subtract the probability (.90) from 1 and divide that result by 2. In this case,

$$(1 - .90)/2 = .05$$

Look up the result (.05) in the table and find its equivalent Z value. (Note: Z will be a negative number.)

$$A = \mu + Z\sigma$$

$$B = \mu - Z\sigma$$

Sampling Distribution of the Mean

Up to now the random variables considered in probability distributions have all been individual measurements. We would now like to consider a distribution in which the random variable is the sample mean. For example, take a sample of five students from a class of 500 and calculate the average GPA. If we take a second sample of five, the average calculated would probably be different from the first. If we continue taking samples of five and continue calculating the average GPA, we would get a distribution of sample means. This distribution of sample means can be treated exactly as another probability distribution. This distribution is one example of a sampling distribution.

The Central Limit Theorem

Note that we are dealing with two different distributions. One is the distribution of original measurements. In our example it is the distribution of the individual GPAs of the 500 students. The second distribution is the distribution of the sample means (n = 5, there would be a different distribution for each different sample size). An important relationship exists between these two distributions. This relationship is called the central limit theorem and is

$$\mu_{\bar{X}} = \mu_X$$

$$\sigma_{\bar{X}} = \frac{\sigma_X}{\sqrt{n}}$$

As n increases in size, the distribution of sample means approaches a normal distribution regardless of the shape of the original distribution of individual measurements. As a rule of thumb, normality can be assumed when $n \geq 30$.

Normal Approximation to the Binomial Distribution

When n is large, calculation of the binomial probability is cumbersome. Also, tables for large n are frequently not readily available. Under these circumstances we find that the normal probability distribution can be used as a reasonable approximation. A rule of thumb suggests that the normal probability distribution is a good approximation if $np > 5$ and $n(1-p) > 5$. The parameters for the normal approximation are

$$\mu = np \qquad \sigma = np(1-p)$$

Here is a table that gives the equivalent normal probability statement that would be used as an approximation for the associated binomial probability statement.

Binomial Probability Statement, Given np	Equivalent Normal Probability Statement, Given $\mu=np$, $\sigma=np(1-p)$
$P(X = A)$	$P(A - .5 \leq X \leq A + .5)$
$P(X \leq A)$	$P(X \leq A + .5)$
$P(X \geq A)$	$P(X \geq A - .5)$
$P(A \leq X \leq B)$	$P(A - .5 \leq X \leq B + .5)$
$P(A \geq X \geq B)$	$P(A + .5 \geq X \geq B - .5)$

This approximation can be extended to proportions (P = proportion success, n = sample size). The parameters for the normal approximation are

$$\mu = p$$

$$\sigma_p = \sqrt{\frac{P(1-P)}{n}}$$

Then when $P\left(\frac{X}{n} \leq P\right)$, $\qquad Z = \dfrac{P - p}{\sqrt{\dfrac{P(1-P)}{n}}}$.

Note that, with proportions, there is no correction factor similar to the binomial approximation.

KEY TERMS

Central limit theorem	Theorem stating that as the sample size becomes large, when each observation is

	selected independently, the distribution of the sample means tends toward a normal distribution.
Correction for continuity	Subtracting ½ interval value from the lowest X value and/or adding ½ interval value to the highest X; used in the normal approximation to the binomial distribution.
Normal curve	Term used to refer to the probability distribution for "normally" distributed variables; the distribution is bell-shaped.
Normal deviate	Value of Z, or the number of standard deviations.
Probability density function	Another term used to describe a continuous probability distribution.
Sampling distribution	The probability distribution of all distinct values of a given sample statistic from all possible samples of equal size taken from the same population.
Sampling distribution of the sample mean	Distribution of the means of samples taken from one population.
Standard error $\sigma_{\bar{X}}$	Standard deviation of the sample mean; standard error of $\bar{X}$.
Standard normal distribution	Probability distribution for the standard normal random variable Z, shown by the standard normal curve: $Z = \frac{(X - \mu)}{\sigma}$.
Z Score	The difference between the observed value of X and its mean in terms of its standard deviation.

FORMULA REVIEW

(8-1) Z Score $Z = \frac{X - \mu}{\sigma}$

(8-2) Mean of $\bar{X}$ $E(X) = \mu_{\bar{X}} = \bar{X}_1 P(\bar{X}_1) + \bar{X}_2 P(\bar{X}_2) + \cdots + \bar{X}_m P(\bar{X}_m)$

(8-3) Mean of the population distribution of X $E(\bar{X}) = E(X)$

(8-4) Variance of $\bar{X}$ $\sigma_{\bar{X}}^2 = \frac{\sigma_X^2}{n}$

(8-5) Standard error of $\bar{X}$ $\qquad \sigma_{\bar{X}} = \dfrac{\sigma_X}{\sqrt{n}}$

(8-6) Standard normal score based on $\bar{X}$ $\qquad Z = \dfrac{\bar{X} - \mu}{\dfrac{\sigma}{\sqrt{n}}}$

(8-7) Z score for bionomial random variable $\qquad Z = \dfrac{X - np}{\sqrt{npq}}$

(8-8) Expected value of P $\qquad E(P) = \left(\dfrac{1}{n}\right)np = p$

(8-9) Variance of P $\qquad \sigma_p^2 = \left(\dfrac{1}{n^2}\right)npq = \dfrac{pq}{n}$

(8-10) Standard deviation of P $\qquad \sigma_p = \sqrt{\dfrac{pq}{n}}$

(8-11) Calculating Z score for P $\qquad Z = \dfrac{P - p}{\sqrt{\dfrac{PQ}{n}}}$

ORGANIZED LEARNING QUIZ

1. The bursting strength of a certain hydraulic hose is known to average 500 pounds with a standard deviation of 5 pounds. It is also approximately normally distributed. What proportion of these hoses will burst only at more than 512 pounds of pressure?

    a. .0152
    b. .0082
    c. .0252

2. What proportion of these hoses (problem 1) will burst at less than 490 pounds?

    a. .0228
    b. .0158
    c. .0318

3. Suppose that 20% of a company's inventoried products are obsolete and that the company has a large number of products. In finding the probability of obsolete products in a large sample, you would use

85

    a. the standard normal distribution.
    b. sampling distribution of a sample mean.
    c. normal approximation to the binomial distribution.

4. Suppose that an accountant in the company in problem 3 randomly selects a sample of ten items. To determine the probability that he will find one or fewer obsolete products, you would

    a. go to the binomial tables.
    b. use the normal approximation.
    c. use the normal table.
    d. There is insufficient information to answer.

5. Suppose that the accountant in problem 4 increases his sample size to 50. What is the probability that he will find 10% or fewer obsolete products in his sample?

    a. less than 3%
    b. approximately 5%
    c. 10%

6. The XYZ Company uses a machine to fill boxes with soap powder. In a state of statistical quality control, the net weights of the boxes of soap are normally distributed with a mean of 15 ounces and a standard deviation of .8 ounce. What proportion of the boxes will have net weights of more than 14 ounces?

    a. .8523
    b. .9654
    c. .8944

7. The average score of 10,000 college graduates who have taken the Graduate Management Admissions Test for graduate school is 500 with a standard deviation of 100. Assuming the data to be normally distributed, if we were to draw a sample of 100 students from this population, what is the likelihood that their average would exceed 520?

    a. 2.3%
    b. 2.7%
    c. 3.2%

8. The kilowatt demand of the area serviced by the Smoke Stack Power Plant is normally distributed with a mean of 120,000 and a standard deviation of 10,000. If the plant can generate at most 150,000 kilowatts, what is the probability that at any given time there will be an overload?

    a. .13%
    b. 1.3%
    c. 13%

9. In working problems using the standard normal distribution, which of the following would *not* be correct?

    a. Thinking that one-half of the area was on one side of the mean and that the other half of the area was on the other side of the mean.

b. Finding the probability that Z >-2.5 by subtracting .0062 from 1,000.
  c. Finding the area between the mean and Z = 2 by doubling the area between the mean and Z = 1.
  d. Finding the probability that Z > 2 by subtracting the probability that Z < 2 from 1.

10. The total area under the normal curve is

    a. .999.
    b. 1.000.
    c. 1.001.
    d. six standard deviations.
    e. infinite.

11. The central limit theorem states that

    a. if n is large, the theoretical sampling distribution of $\bar{X}$ can be approximated very closely with a normal curve.
    b. if n is small, the standard error of the mean is small.
    c. if $\bar{X}$ is small, the standard error of the mean is small.
    d. a normal curve does not provide a good fit if it has central limits.
    e. the standard deviation and the standard error of the mean must be equal in a normal distribution.

12. Each score in a distribution is multiplied by 4. The mean will be _____ times the old mean, the variance will be _____ times the old variance, and the standard deviation _____ times the old standard deviation.

13. What percentage of a normal distribution is within plus and minus one σ of the mean? _____

14. In a normal curve, _____ % of the area lies between -2σ and $+1\sigma$. _____ % of the normal curve lies between $\pm 2\sigma$.

15. If X = 8, $\bar{X}$ = 4, and S = 2, Z = _____.

16. If your score on a test is below the class mean, your Z score will be (positive/negative/zero). If your score is above the class mean, your Z score will be (positive/negative/zero). If your score is equal to the class mean, your Z score will be _____.

17. The percentage of the between Z scores of -1.0 and -2.0 is _____.

18. If your raw score is 10, the class mean is 11, and the standard deviation is 5, your Z score is _____. You did better than _____ % of your group. _____ % did better than you.

19. Other things being equal, we expect a sample of size 300 to be _____ (more/less) like the population than a sample size of 40.

20. The lifetime of a particular model of a dry cell battery is normally distributed with a mean of μ = 1000 hours and a standard deviation of σ = 100 hours. Find the probability

that one of these batteries will last

    a. between 1000 and 1150 hours.
    b. between 950 and 1000 hours
    c. more than 1250 hours.

21. What is a normal probability distribution?

   _____
   _____
   _____

22. Define the conditions necessary for a probability distribution to be a binomial probability distribution.

   _____
   _____
   _____

23. In the area below, sketch a normal probability curve and indicate on the curve $P(X \geq 80)$. Assume that $\mu = 50$ and $\sigma = 10$.

24. In computing probabilities with the normal distribution, the Z value or standard normal deviate is calculated as $Z = (X - \mu)/\sigma$. Why is this necessary?

   _____
   _____
   _____

25. Explain the relationship between X and $\bar{X}$.

   _____
   _____
   _____

ANSWERS TO ORGANIZED LEARNING QUIZ

1.b. $Z = \dfrac{512 - 500}{5} = 2.4$; from table, $Z(2.4) = .4918$

    $.5000 - .4918 = .0082$

2. a. $Z = \dfrac{490 - 500}{5} = -2$; from table, $Z(-2) = .4772$

   $.5000 - .4772 = .0228$

3. c.

4. a.

5. b. $Z = \dfrac{X - np}{\sqrt{npq}} = \dfrac{5.5 - 10}{\sqrt{50(.2)(.8)}} = \dfrac{-4.5}{2.83} = -1.59$; from table, $Z(1.59) = .44408$; $.5000 - .44408 = .05592$

6. c. $Z = \dfrac{X - \mu}{\sigma} = \dfrac{14 - 15}{.8} = \dfrac{1}{.8} = 1.25$; from table, $Z(1.25) = .3944 + .5000 = .8944$

7. a. $Z = \dfrac{\bar{X} - \mu}{\sigma_{\bar{X}}} = \dfrac{520 - 500}{100/\sqrt{100}} = \dfrac{20}{10} = 2$

   $Z(2) = .4772$; $.5000 - .4772 = .0228$

8. a. $Z = \dfrac{150{,}000 - 120{,}000}{10{,}000} = 3$; from the table, $Z(3) = .0013$

9. c.

10. b.

11. a.

12. 4, 16, 4

13. 68.27

14. 81, 95

15. 2

16. negative, positive, zero

17. 13.59

18. -.2, 42%, 58%

19. More

20. a. .4332
    b. .1915
    c. .2420

21. It is a listing of the probabilities for every possible value of a random variable distributed symmetrically about the mean. A frequency distribution approximated by the shape of a normal curve is normally distributed.

22. Most have only two possible outcomes; probability of success remains constant from trial to trial; trials are independent of each other.

23.

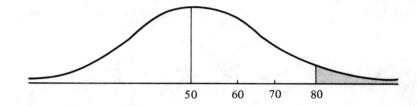

24. To convert from raw data scale to the Z scale to allow the use of the normal distribution table.

25. $\bar{X}$ is a random variable, the values of which are the means of all possible samples drawn from the population distribution of X; $\mu_X = \mu_{\bar{X}}$.

SAMPLE EXERCISE

An internal auditing department has the responsibility of periodically evaluating the control procedures in the accounts payable department as well as approving the monthly invoices for payment. One step in the process of evaluation is to sample the invoices on a month-by-month basis. The auditing department sets an acceptable level of error; if the sample of invoices results in an error rate less than or equal to the acceptable level of error, the invoices for the period in question are approved for payment. If the number of errors exceeds the acceptable level, additional checking measures are used. Since these measures involve additional expense, it is of obvious benefit to the company to have procedures that minimize the risk of error while minimizing the costs of processing invoices.

The table below shows the results of sampling over a 13-week period, with the descriptive statistics calculated.

Week No.	X*	X^2
1	.032	.001024
2	.020	.0004
3	.015	.000225
4	.035	.001225
5	.028	.000784
6	.012	.000144
7	.019	.000361
8	.027	.000729
9	.025	.006625
10	.030	.0009
11	.027	.000729
12	.028	.000784
13	.035	.001225

*Errors as a percent of the total number of errors possible.

Totals   ($\bar{X}$)   Ⓐ .333   Ⓑ .00938
                         Ⓒ .0256   Ⓓ .00853
                                           Ⓔ .00085
                                           Ⓕ .000071
                                           Ⓖ .00842

Explanation of Table

Ⓐ   Sum of the observations
Ⓑ   Sum of the squares
Ⓒ   Arithmetic mean
Ⓓ   Correction (Ⓐ * Ⓒ)
Ⓔ   Sum of the squares of the deviations (Ⓑ - Ⓓ)
Ⓕ   Sample variance (Ⓔ/12)
Ⓖ   Sample standard deviation ($\sqrt{Ⓕ}$)

     Assume that the error rate is normally distributed with a mean (μ) of .025 and a standard deviation of .010. The sample of 13 weeks is shown in the table and the calculations made. What is the probability that $\bar{X}$ will be greater than .030?

     The desired probability here is $P(\bar{X} > .030)$. With a standard deviation $\sigma = .010$ and the sample size = 13, we can convert $\bar{X}$ to a Z score by Eq. (8-6). The probability asked for is

$$P(\bar{X} > .30) = P\left(Z > \frac{.030 - .025}{.1/\sqrt{13}}\right) = P\left(Z > \frac{.005}{.0277}\right) = P(Z > .1805)$$

$$= 1 - P(Z < .1805)$$

$$= 1 - .5714 = .4286.$$

In other words, there is a fairly high probability (43%) that the average number of errors will exceed .030.

     In the samples shown in the table, all the error rates fall between -2σ and $+1\sigma$. The department would primarily be concerned with the upper end of the distribution--in other words, maintaining the error rate at or below some level. If the acceptable error rate is set at .035, the probability of exceeding that level is 36%.

$$P(\bar{X} > .035) = P\left(Z > \frac{.035 - .025}{.1/\sqrt{13}}\right) = P\left(Z > \frac{.010}{.0277}\right)$$

$$= P(Z > .3610)$$

$$= 1 - .6406 = .3594$$

CASE PROBLEM

     Wheeling and Dealing Alfalfa Company buys hay from a number of different farms and sells the hay to a feed processor. Lee Shindig, the manager of W & D, has located 10 or 12 farms in the area, each of which has three or four stacks of hay of varying quality. In this instance, quality is measured primarily by protein content. When Lee delivers the hay to the processors, the protein content is calculated. If the protein content falls below 15% for any given load, the processors reduce the amount that they are willing to pay by $6 per ton.

Because of the importance of protein content, Lee regularly takes samples from the various stacks on each farm and has them tested by a local analysis laboratory. The results from the analysis of the most recent set of samples is shown below. The lot number is the number that Lee has assigned to each haystack for reference purposes. It will be seen that Lee took two samples from lots 1 through 7 and one sample each from lots 8 and 9. "Sample Source" shows the farm location of the lot and the volume of hay located in the lot.

Table of Sample Results
Wheeling and Dealing Alfalfa Company

Lot No.	Number	Sample Sources	Sample Protein Content	Lot Protein Average
1	1	Farm #8	14.4	
	2	275 tons	17.7	16.05
2	3	Farm #4	15.1	
	4	415 tons	17.4	16.25
3	5	Farm #5	14.6	
	6	235 tons	14.7	14.65
4	7	Farm #6	14.9	
	8	565 tons	16.5	15.00
5	9	Farm #7	14.9	
	10	160 tons	16.4	15.65
6	11	Farm #3	14.3	
	12	295 tons	16.5	15.40
7	13	Farm #9	13.9	
	14	330 tons	16.3	15.10
8	15	Farm #2	15.9	15.90
		96 tons		
9	16	Farm #1	15.6	15.60
		345 tons		
	Total	2716 tons		

Answer the following questions, using the table information.

1. What is the mean protein average across all lots?

2. What is the standard deviation across all lots?

3. Assuming that Lee buys the entire 2716 tons, how many tons would you anticipate will fall below the 15% protein level?

4. Of the nine lots sampled, which lot is most likely to show a protein deficiency? What is the probability that this lot will, in fact, show a protein deficiency?

5. What is the probability that lot No. 4 will show a protein deficiency?

9

STATISTICAL ESTIMATION

CHAPTER LEARNING OBJECTIVES

On completing this chapter, the student should be able to

- Describe what is meant by statistical estimation.

- Define the terms point estimate and confidence interval.

- Identify the three criteria for a good estimator.

- Calculate confidence intervals for the mean.

- Calculate confidence intervals for the difference of two means.

- Calculate confidence intervals for proportions.

- Calculate confidence intervals for the difference of two proportions.

- Identify assumptions inherent in the preceding confidence intervals.

- Determine the sample size required for a specified level of accuracy for confidence intervals relating to the mean and proportion.

SYNOPSIS

So far you have developed some ideas on calculating statistics, such as the mean and the standard deviation. You have also developed an understanding of probability and probability distributions and have been exposed to ideas concerning sampling and sampling distributions. All this provides the necessary background to undertake the study of inferential statistics. Inferential statistics probably represents the primary use of statistics. It involves the question of taking a sample from a population and then drawing conclusions about the population from the sample. Inferential statistics can be divided into two major areas. The first to be considered is statistical estimation. Hypothesis testing, the other major area, will be discussed in later chapters.

As mentioned, we are frequently faced with the problem of drawing conclusions about a population when it is practically or economically not feasible to examine the total population. Our procedure is to examine a part of the population from the sample. If the statement that we wish to make is in the form of an estimate of population parameters, we are then involved with statistical estimation. Basically statistical estimation concerns the relation between statistics, summary properties of the sample, and parameters, summary properties of the population.

Statistics (in particular, the sample mean, sample variance, and sample proportion) is the summary measures of the sample and the values that we will use to estimate the equivalent population parameters. In order that a statistic be considered a good estimator, it should possess certain properties. It should be consistent; that is, it should be closer to the true value of the population as the sample size gets larger. A statistic should be efficient. When compared to other alternatives for estimating the same population parameter, it should have the smallest standard error. A statistic should also be unbiased; that is, the expected value of the sample statistic should equal the population parameter. The primary statistics of interest in this chapter are the sample mean ($\bar{X}$), the sample variance (S^2), and the sample proportion (P). All meet the criteria specified for good estimators.

Point Estimation

One type of estimate used is called a point estimate. It is the single value that represents the best knowledge of the population parameter. The values of the sample statistics are the best point estimates. The sample mean ($\bar{X}$), the sample standard deviation (S) or the variance (S^2), the sample proportion (P), the difference between sample means ($\bar{X}_1 - \bar{X}_2$), and the difference of two proportions ($P_1 - P_2$) are the sample statistics used in this chapter that also represent the point estimate or best knowledge of their equivalent population parameters.

Interval Estimations about Means

The second type of estimate is called an interval estimate. An interval estimate not only reflects the point estimate of the population parameter; it also provides an indication of our uncertainty about that estimate. Most of us have frequently used interval estimates on an intuitive basis. For example, you may have estimated the gas mileage for you car as between 15 and 20 miles per gallon. This is an interval estimate. You aren't certain of the exact value, but you are reasonably sure that the true value lies somewhere in the interval. This is exactly the sense of what is meant by an interval estimate in statistics. The chief difference is that instead of saying we are reasonably certain that the true value falls in the interval, we will state a probability that the true value lies in the interval. This probability, termed the *level of confidence*, will be represented by the expression $1 - \alpha$. α is the symbol that will be used to represent the probability that the confidence interval will not contain the true value of the statistic or it is the probability that the interval is in error.

It is true that the probability that Z lies between -1.96 and 1.96--$P(-1.96 \leq Z \leq 1.96)$--is .95. It can be confirmed by referring to the normal probability tables. If we substitute $Z = \frac{(\bar{X} - \mu)}{\sqrt{n}}$ and then $-1.96 \leq \frac{(\bar{X} - \mu)}{\sigma/\sqrt{n}} \leq 1.96$, all terms can be multiplied by $\sigma/\sqrt{n}$, resulting in $-1.96\sigma/\sqrt{n} \leq \bar{X} - \mu \leq 1.96\sigma/\sqrt{n}$. Finally, the sample mean is subtracted from all terms and all terms are then multiplied by a -1, resulting in $\bar{X} - 1.96\sigma/\sqrt{n} \leq \mu \leq \bar{X} + 1.96\sigma/\sqrt{n}$. This says that 95% of the time the true mean will lie within $\pm 1.96\sigma/\sqrt{n}$ of the sample mean. In this case, α equals .05 and $1 - \alpha$, the confidence interval, equals .95. Rearranging terms and generalizing the result yields the expression for the confidence intervals for means.
$$\bar{X} \pm Z_{(\alpha/2)} \sigma/\sqrt{n}$$

Interval About Proportions

Let us examine the confidence interval, term by term, for the general form is similar for most confidence intervals. The first part of the confidence interval, $\bar{X}$, is the point estimate of the statistic of interest. If could have been $\bar{X}_1 - \bar{X}_2$, P, $P_1 - P_2$, or other possibilities that will be taken up in later chapters. This point estimate represents the most likely true value of the population parameter, given the information available. The second part of the expression, $Z_{(\alpha/2)} \sigma/\sqrt{n}$ represents the uncertainty or potential error in the estimate. The $Z_{(\alpha/2)}$ reflects the assumption made about the probability distribution and the confidence interval $(1 - \alpha)$ that has been chosen. Use of a Z value relfects the assumption that the statistic is distributed according to a normal distribution. It also reflects the assumption that we know the population standard deviation, in the case of intervals concerning means, or that the sample size is reasonably large in the case of proportions. The $\sigma/\sqrt{n}$ represents the standard error related to the sample statistic. In the case of estimating a mean, the standard error is $\sigma_{\bar{X}} = \sigma/\sqrt{n}$. In the case of estimating proportions, the standard error is $\sigma_P = \sqrt{\frac{P(1 - P)}{n}}$. When dealing with the difference of means, the standard error becomes
$$\sigma_{\bar{D}} = \sqrt{\frac{\sigma_1^2}{n_1} + \frac{\sigma_2^2}{n_2}}$$

The standard error of the difference of proportions is
$$\sigma_{P_1 - P_2} = \sqrt{\frac{P_1 Q_1}{n_1} + \frac{P_2 Q_2}{n_2}}$$

In summary, our confidence interval is usually made up of three parts-- the point estimate or sample statistic, the tabulated value from the statistical distribution for the confidence interval, and the standard error associated with the sample statistic. Thus in generating a confidence interval, we first identify the sample statistic to be used. Associated with any sample statistic is a standard error. Finally, we select the confidence interval $(1 - \alpha)$ and identify, from assumptions the appropriate statistical distribution. Then we look up the tabulated value for the probability associated with the confidence interval. All these go back together in the general form:

Sample statistic ± the tabulated value for confidence interval x the standard error.

Determination of Sample Size

The error or spread of the confidence interval can be influenced in two ways by the user. Selection of α has a direct bearing on the interval size. A smaller value of α leads to a larger confidence interval and a larger value of α leads to a smaller confidence interval. This situation is true because the magnitude of Z is inversely related to the magnitude of α.

The other source of control over the confidence interval is the sample size. In examining the formula for the confidence interval, it should be realized that the standard error directly affects the size of the confidence interval. Note that in all types of standard errors used in this chapter the sample size is in the denominator of the formula. Thus as sample size increases, the standard error gets smaller.

The expression $Z_{(\alpha/2)} \sigma/\sqrt{n}$ represents the half-width of the spread of the confidence interval. In this sense, it is often referred to as the *maximum error* of the estimate, which is expressed as

$$e = Z_{(\alpha/2)} \sigma/\sqrt{n}.$$

It is possible to solve that expression for n, the sample size using

$$n = \left(Z_{(\alpha/2)}\right)^2 \sigma^2 / e^2.$$

Given a value for the population standard deviation, the value for the level of confidence (1 - α) and its associated Z value, it is then possible to solve for n, the sample size for a stated or given error or spread.

Similarly, it is possible to solve for the necessary sample size for a given maximum error when estimating a confidence interval for a proportion.

$$n = p(1 - p)(Z_{\alpha/2}/e)^2$$

In this case, it is necessary to make some rough estimate of p or to assume the worst case of p = .5.

KEY TERMS

Confidence interval — Probability that the interval obtained for estimation of the population parameter contains the true mean. Measures the

	probability of error. Therefore $1 - \alpha$ is the confidence interval.
Confidence limits	Upper and lower limits of the confidence interval.
Consistent estimator	Sample statistic that comes closer to the value of the estimated population parameter as the sample size gets larger.
Efficient estimator	Estimator that has the smaller standard deviation when compared to another estimator.
Interval estimation of the mean	Procedure of determining an interval of values that will include the true value of a population mean with a certain probability.
Point estimate	Value resulting from the use of a point estimator.
Point estimation	If a single value calculated from the observations in a sample is used as an estimate of an unknown population parameter, the procedure is referred to as point estimation.
Point estimator	Rule or method that tells us how to calculate a single value from sample data.
Sampling error	Error that occurs when a sample statistic is used to estimate a population parameter; the distance between the value of the estimator and the value of the estimated parameter.
Unbiased estimator	Sample statistic that has a mean equal to the parameter of the population from which the sample is taken.

FORMULA REVIEW

Confidence interval for means

$$\bar{X} \pm Z_{(\alpha/2)} \sigma/\sqrt{n}$$

Assumes population standard deviation is known.

Confidence interval for difference of two means

$$\bar{D} \pm Z_{(\alpha/2)} \sqrt{\frac{\sigma_1^2}{n_1} + \frac{\sigma_2^2}{n_2}}$$

$$\bar{D} = \bar{X}_1 - \bar{X}_2$$

Assumes population standard deviation is known.

Confidence interval $\quad P \pm Z_{(\alpha/2)} \sqrt{\dfrac{P(1-P)}{n}}$

P = sample proportion

$S_p = \sqrt{\dfrac{P(1-P)}{n}}$

Confidence interval for the difference between two proportions $\quad (P_1 - P_2) \pm Z_{(\alpha/2)} \sqrt{\dfrac{P_1 Q_1}{n_1} + \dfrac{P_2 Q_2}{n_2}}$

Sample size determination estimating means $\quad n = \dfrac{(Z_{\{\alpha/2\}})^2 \sigma^2}{e^2}$

e = sampling error or maximum error of the estimate

Sample size determination estimating proportions $\quad n = p(1-p)\left(\dfrac{Z_{(\alpha/2)}}{e}\right)^2$

ORGANIZED LEARNING QUIZ

1. A candy company wishes to estimate the average daily output, in pounds, of its chocolate-processing machine. The weekly yields are known to have a variance of 7 pounds. The record for 49 days shows an average yield of 100 pounds/week. The estimated average weekly yield is

    a. 107.
    b. 100.
    c. unknown; insufficient information to make an estimate.

2. In problem 1 the possible sampling error is

    a. within three standard deviations, or ±3 pounds.
    b. within three standard deviations, or ±7 pounds.
    c. unknown; insufficient information to make an estimate.

3. The average score of 81 senior business majors at Broken Bow State College who took the Graduate Management Admission Test is 510. If the standard deviation of scores is 36, the 95% confidence limit for the true mean of the test scores is

    a. 510 ± 36.
    b. 510 ± 6.0.
    c. 510 ± 7.84.

4. The mean education level of employees in 49 import industries was found to be 11.453 with a variance of 1.05; the mean education level in 49 export industries was found to be 11.960 with a variance of .55. The confidence interval for the true difference between the true means is ($\alpha = .05$)

   a. -.056 to +.056.
   b. -.456 to +.963.
   c. -.430 to +1.44.
   d. -.254 to +.7605

5. Suppose that a researcher wishes to know the true proportion of the workforce in a particular industry that is female. A sample of 81 industries is selected with the result that P=.43. The 98% confidence interval for P, the true proportion of the workforce that is female, is

   a. .302 to .558.
   b. .40 to .46.
   c. .366 to .494

6. A Steve Grieve poster is to be tested for its effect on blood pressure. Twelve women have their blood pressure tested before and after the stimulus. The results are

	Before	After
Sample size	$n_1 = 12$	$n_2 = 12$
Sample mean	$\bar{X}_1 = 127.9$	$\bar{X}_2 = 135$
Population variance	$\sigma_1^2 = 5.3$	$\sigma_2^2 = 12.7$

   The 95% confidence interval for the true difference between the true means is

   a. $-9.0 < \delta < -5.0$.
   b. $-9.49 < \delta < -4.71$.
   c. $-12.4 < \delta < +5.6$.

7. In a prereferendum poll of 625 voters in a campaign to eliminate nonreturnable beverage containers, 45% of the voters were opposed. The 95% confidence interval for the true proportion opposed to the container control bill is

   a. .411 to .489.
   b. .405 to .495.
   c. .4275 to .4725.

8. An operator of a fast-food franchise wishes to estimate the average number of customers served per hour by a certain employee. He insists that the error of estimation will be no more than seven customers with a probability of .95. Assume that the population standard deviation is known to be 15. The sample size needed to meet the operator's requirements is

a. 14.
b. 16.
c. 18.

9. A service company is considering switching to flexible hours for employees. If 70% are in favor, the company will make the change. The president estimates that, in fact, only 60% will be in favor. The sample size needed to check the president's assumption with no more than 5% error and 90% confidence is

a. 215.
b. 260.
c. 295.

10. Under the Equal Pay Act companies are required to pay women equal pay if they are performing essentially the same work as men. The manager of Marvin's Car Products wishes to determine if the company is in compliance with the law and compiled the following information on people working in a particular kind of job.

	Male	Female
Sample size	$n_1 = 75$	$n_2 = 50$
Sample mean (wage)	$\bar{X}_1 = \$5.$	$\bar{X}_2 = \$3.50$
Variance	$\sigma_1^2 = \$1.25$	$\sigma_2^2 = 2.50$

The 95% confidence interval for the true difference between male and female wages is

a. .75 ± .51.
b. 1.50 ± .51.
c. 1.50 ± .76.

11. The range that we estimate with a given probability will include the true population parameter is called the _____ _____.

12. The size of the sample needed varies proportionally with the size of the universe from which the sample was taken.

a. True
b. False

13. If we increase the degree of certainty, the confidence interval becomes (wider/narrower).

14. To increase the degree of certainty, we may make the sample size (larger/smaller).

15. The greater the degree of desired certainty, the (larger/smaller) the α ratio and the (larger/smaller) the Z value.

16. The 95% confidence interval for the true mean when $\bar{X} = 25$, $\sigma = 5$, and $n = 25$ is 25 ± _____.

17. The 90% confidence interval for the true proportion given P = .50, and N = 25 is .5 ± _____.

18. A single value calculated from the observations in a sample to estimate an unknown population parameter is called a _____ _____.

19. The distance between the value of an estimator and the true value of the parameter is called the _____ _____.

20. If N is _____, substituting P for p in p(1 - 1)/n will not distort the results appreciably.

21. Name the three criteria used to evaluate the quality of a point estimator.

    _____
    _____
    _____

22–24. In problems 22 to 24 define the criteria listed in problem 21.

22. _____
    _____
    _____

23. _____
    _____
    _____

24. _____
    _____
    _____

25. Other than the cost of obtaining information, three determinants of sample size are discussed in the text. List and describe these three.

    _____
    _____
    _____

ANSWERS TO ORGANIZED LEARNING QUIZ

1. b. $\bar{X}$ is a point estimator of μ.

2. a. $\sigma_{\bar{X}} = \dfrac{7}{\sqrt{49}} = \dfrac{7}{7} = 1;$

   $3(1) = 3;$ the true average will probably be between 97 and 103 pounds.

3. c. $\bar{X} - Z_{\alpha/2} \dfrac{\sigma}{\sqrt{n}} < \mu < \bar{X} + Z_{\alpha/2} \dfrac{\sigma}{\sqrt{n}}$

   $Z_{\alpha/2} = Z_{.05/2} = Z_{.025} = 1.96$

   $\sigma_{\bar{X}} = \dfrac{\sigma}{\sqrt{n}} = \dfrac{36}{\sqrt{81}} = \dfrac{36}{9} = 4$

   $510 - 1.96(4) < \mu < 510 + 1.96(4)$

   $510 - 7.84 < \mu < 510 + 7.84$

   There is a 95% probability that the interval 502.16 to 517.84 will include the true mean.

4. c. $\bar{D} - Z_{\alpha/2} \sqrt{\dfrac{\sigma_1^2}{n_1} + \dfrac{\sigma_2^2}{n_2}} < \delta <$

   $\bar{D} + Z_{\alpha/2} \sqrt{\dfrac{\sigma_1^2}{n_1} + \dfrac{\sigma_2^2}{n_2}}$

   $(11.453 - 11.960) - 1.96 \sqrt{\dfrac{1.05}{49} + \dfrac{.55}{49}} < \delta <$

   $(11.453 - 11.960) + 1.96 \sqrt{\dfrac{1.05}{49} + \dfrac{.55}{49}}$

   $(-.507) - 1.96(.478) < \delta < (.507) + 1.96(4.78)$

   $-.507 - .9371 < \delta < -.507 + .9371$

   $-1.4441 < \delta < .4301$

5. a. $(.43) - 2.33\sqrt{\frac{(.43)(.57)}{81}} < p < (.43) + 2.33\sqrt{\frac{.43(.57)}{81}}$

   $.43 - 2.33(.055) < p < .43 + 2.33(.055)$

   $.43 - .128 < p < .43 + .128$

   $.302 < p < .558$

6. b. $-7.1 - 1.96\sqrt{\frac{5.3}{12} + \frac{12.7}{12}} < \delta <$

   $-7.1 + 1.96\sqrt{\frac{5.3}{12} + \frac{12.7}{12}}$

   $-7.1 - 1.96(1.22 < \delta < -7.1 + 1.96(1.22)$

   $-7.1 - 2.39 < \delta < -7.1 + 2.39$

   $-9.49 < \delta < -4.71$

7. a. $.45 - 1.96\sqrt{\frac{.45(.55)}{625}} < p < .45 + 1.96\sqrt{\frac{.45(.55)}{625}}$

   $.45 - .039 < p < .45 + .039$

   $.411 < p < .489$

8. c. $1.96\sigma_{\bar{X}} = 7$ or $1.96\frac{\sigma}{\sqrt{n}} = 7$

   $1.96(15) = 5\sqrt{n}$

   $\sqrt{n} = \frac{1.96(15)}{7}$ and $n = 4.2^2 = 17.64$ or $\cong 18$

9. b. $1.6455\, S_p = .05$

   $S_p = \sqrt{\frac{.6(.4)}{n}}$; $1.645\sqrt{\frac{.6(.4)}{n}} = .05$

   $1.645(.49) = .05\sqrt{n}$

   $\sqrt{n} = \frac{.806}{.05} = 16.12$

   $n = 259.85$ or approximately 260

10. b. $1.50 \pm 1.96\sqrt{\frac{1.25}{75} + \frac{2.50}{50}}$

$1.50 \pm 1.96\sqrt{.0667}$

$1.50 \pm .51 = .99 < \delta < 2.01$

11. confidence interval
12. b. False
13. wider
14. larger
15. smaller, larger
16. 1.96
17. (.1)1.645
18. point estimator
19. sampling error
20. large

21. Consistency, efficiency, lack of bias (relative)

22-24. See Key Terms section of this chapter.

25. Accuracy: refers to the level of sampling error

Confidence: refers to the extent to which the interval estimated contains the true population parameter

Standard deviation: range from lowest to highest values divided by six may be used as an approximation, since $\mu \pm 3$ standard deviations will include almost 100% of all potential observations.

SAMPLE EXERCISE

Ms. I.M. Precise, a systems analyst in a billing department, is interested in the error rate associated with the billing form currently being used. She feels that an improved form, while costing more, might reduce the error rate sufficiently to cover the additional cost and result in substantial savings. She wishes to use the new form on a trial basis and compare the results.

1. How large a sample is necessary to allow a satisfactory comparison? She specifies that the error of estimation be no more than .05; she assumes that the new proportion of errors will be about 25% below the old rate, which was .02.

$n = \frac{(Z_{\alpha/2})^2 \sigma^2}{e^2}$; $1 - \alpha = .9$; $\frac{\alpha}{2} = .05$

$Z(.05) = 1.645$

error specified at .05

$1.645 S_p = .05$

estimated value of p = .015   (25% below old value of .02)

standard error is

$$S_p = \sqrt{\frac{(.015)(.985)}{n}}$$

$$1.645\sqrt{\frac{(.015)(.985)}{n}} = .05$$

$$1.645(.122) = .05\sqrt{n}$$

$$\sqrt{n} = \frac{1.645(.122)}{.05}$$

$$n = \frac{(.20069)^2}{(.05)^2} = \frac{.04028}{.0025} \cong 16$$

2. The systems analyst decides to try out the new form for 16 weeks and compare the results with the same 16-week period for the previous year. At the end of the trial period she makes up the table below

	Old Form	New Form
Sample size	$n_1 = 16$	$n_2 = 16$
Sample mean	$\bar{X}_1 = .02$	$\bar{X}_2 = .012$ (mean errors)
Population Variance	$\sigma_1^2 = .005$	$\sigma_2^2 = .010$ (assume that these are known)

Ms. Precise decides to construct a 95% confidence interval for the true difference in error rates between the old and new forms.

1. From Eq. (9-5) in the text,

$$(.02 - .012) \pm 1.96\sqrt{\frac{.005}{16} + \frac{.010}{16}}$$

$$.008 \pm 1.96(.031)$$

$$.008 \pm .06$$

2. The confidence interval becomes $-.052 < \delta < +.068$. In other words, the systems analyst can feel 95% certain that the true difference between the difference in error rates is between -.052 (the error rate for the old form is below that of the new form) and +.068 (the rate for the new form is lower).

CASE PROBLEM

A sales manager wishes to compare the performance of two

salesmen over a 10-week period. The salesmen sell nine different items. The results are shown below.

               Ten-Week Averages (Means)

Product	Salesman I	Salesman II
1	13	18
2	19	13
3	14	17
4	12	17
5	21	16
6	14	19
7	17	22
8	20	24
9	17	19

Assume that the sales manager knows the population variances to be $\sigma_1^2 = 3$ and $\sigma_2^2 = 6$.

Construct a confidence interval for the difference in average sales performance between the two salesmen.

10

HYPOTHESIS TESTING

CHAPTER LEARNING OBJECTIVES

Upon completion of this chapter the student should be able to

- Describe and execute the steps in hypothesis testing.

- Formulate the null and alternative hypothesis.

- Distinguish between a one-tailed or two-tailed test.

- Identify and define the possible errors (type I and type II) in hypothesis testing.

- Carry out the test of a hypothesis for means, proportions, difference of two means, and difference of two proportions.

SYNOPSIS

The preceding chapter dealt with the area of statistical estimation, which was the first introduction to statistical inference. This chapter continues the idea of statistical inference--the drawing of conclusions about a population based on a sample. Statistical estimation involves making estimates about the true value of population parameters, based on sample statistics. Hypothesis testing takes a somewhat different approach. In hypothesis testing alternative statements are made about the population parameters and then, on the basis of evidence from the sample statistics, we attempt to decide which statement seems to be true. In reaching this decision, we rely on probability as a guide. The two statements that will be made about the population are termed the null hypothesis and the alternative hypothesis.

Essential Steps of Hypothesis Testing

To understand hypothesis testing fully, it is critical to know the role of the null and the alternative hypotheses. The general point is to assume that the null hypothesis is true unless we have substantial evidence to reject the idea; then we accept the alternative hypothesis. The question of what is substantial evidence involves the question of probability. The probability of the sample result, given the null hypothesis, can be calculated by using the methods described earlier. If the probability is lower than what we are willing to accept, we then choose to accept the alternative hypothesis as the true statement about the population. Accepting the alternative hypothesis thus occurs only when we are quite certain that it is true.

This discussion leads to some general guidelines for formulating hypothesis statements. One approach suggests that we should state in the alternative hypothesis the statement that we wish to prove true. The null hypothesis then becomes the opposite. If testing a new drug, we are interested in proving that the drug is effective. This, then, becomes the alternative hypothesis, whereas the null hypothesis would state that the drug was not effective. A variation would suggest that the alternative hypothesis represents a statement about the population that would be accepted only if it is quite certain that it is true. In the earlier example we do not want to accept the idea that the drug is effective unless we are quite certain that it is. This results in the same null and alternative hypothesis. If we are testing the quality of a product, generally we would assume that the quality was good. We only want to conclude that quality is bad when we are quite certain that the conclusion is true because the conclusion of bad quality usually results in remedial measures and added expense. In this case, our alternative hypothesis is that the quality is bad. The null hypothesis is that the quality is good.

In actual hypothesis testing the probability of the sample result is usually not calculated directly. Probability enters into hypothesis testing through establishing allowable limits on the probability of error. In testing a hypothesis, two types of error can occur. The null hypothesis can be rejected when it is true. This is termed a type I error. The other type of error occurs if the null hypothesis is accepted when it is not true--this is a type II error. The symbol α represents the probability of a type I error and is established by the user prior to testing the hypothesis. The symbol β represents the probability of a type II error.

Several general steps should be followed in carrying out hypothesis testing.

1. Identification of the pattern of the population distribution.
2. Formulation of the null and alternative hypotheses.
3. Specification of the significance level.
4. Formulation of the decision rule.
5. Decision making: accepting or rejecting the null hypothesis.

Selection of a probability distribution involves making key assumptions about the form of the data. Up to now we have dealt with normal probability distributions. The central limit theorem gives powerful support to this assumption. In later chapters we will deal with distributions other than the normal one. The selection of the particular distribution is important to the validity of the conclusions finally reached.

Formulation of the hypothesis was treated earlier in this discussion. What we wish to prove true or what we wish to accept--only if we are quite certain it is true--is formulated as the alternative hypothesis. The null hypothesis then becomes the opposite.

The level of significance, α, is the value for the probability of a type I error. It is assigned by the user prior to the test.

The type I error is then used in the next step of establishing the decision rule. Using the assumption of the probability distribution and the hypothesized value for the test statistic, we calculate or look up in a table the value of the test statistic that represents the type I error. This becomes the critical value and serves to define the critical region where the null hypothesis would be rejected. For the normal distribution, the Z value that is equivalent to the value of $\alpha/2$ and $1 - \alpha/2$ is calculated for a two-tailed test or to the value of α for a one-tailed test (or $1 - \alpha$, depending in which tail the test is). This Z value, the critical value, defines the critical region where the null hypothesis will be rejected.

Finally, the sample is taken and the sample statistic calculated. If the sample statistic falls in the critical region, the null hypothesis is rejected in favor of the alternative hypothesis. If the sample statistic does not fall in the critical region, the null hypothesis is accepted. Thus far discussion has covered hypothesis testing for the sample mean. It has served to illustrate the general nature of hypothesis testing. In this chapter hypothesis testing for other statistics is also treated. Included are the difference of two means, the sample proportion, and the difference of two proportions. All tests assume a normal distribution. The tests concerning the mean and difference of two means require that the populations' standard deviation be known. The test on proportions and the difference of sample proportions requires that the sample size be sufficiently large (np, nq > 5). The formulas for the calculation of the appropriate test statistic, Z, are given in the following formula review. All hypothesis tests in this chapter follow the general procedure outlined for the test for the sample mean.

KEY TERMS

Alternate hypothesis	H_1: tentative statement that a population parameter has a different value than the one specified in the null hypothesis.
Critical region	Another term for the rejection region or that region which will lead us to reject the null hypothesis.
Critical value	Value that separates the rejection from the nonrejection regions.
Decision making	The process of making choices among alternatives.
Decision rule	Rule that specifies the critical region.

Exact hypothesis	Hypothesis that specifies a unique value for the population parameter.
Hypothesis	Tentative statements about one or more population parameters.
Hypothesis testing	Procedure of decision making through proper statistical tests between two competing hypotheses.
Inexact hypothesis	Hypothesis that specifies that a population parameter may assume any one of a set of values.
Level of significance	α: probability of rejecting a true null hypothesis.
Nonparametric approach	Distribution-free approach to hypothesis testing that requires no specification of the population distribution.
Nonrejection region	Region of values for the test statistic that will require that we not reject the null hypothesis.
Null hypothesis	H_o: tentative statement formed primarily to determine whether it can be rejected.
One-tailed test	Test required when the alternative hypothesis does specify the direction of the difference.
Parametric approach	Approach to hypothesis testing that requires the identification of the probability distribution being examined.
Power of the statistical test	$1 - \beta$: probability that a statistical test will lead to the correct decision with respect to rejecting the null hypothesis when it, in fact, is false.
Rejection region	Region of values for the test statistic that will lead us to reject the null hypothesis.
Test statistic	Random variable, the value for which is used for the decision to reject or not reject the null hypothesis.
Two-tailed test	Test required when the alternative hypothesis does not specify the direction of the difference.
Type I error	Occurs when a true null hypothesis is rejected.
Type II error	Occurs when we fail to reject a false null hypothesis.

FORMULA REVIEW

Test for means

$$Z = \frac{\bar{X} - \mu}{\sigma/\sqrt{n}}$$

Assume σ is known.

Test for the difference of two means

$$Z = \frac{(\bar{X}_1 - \bar{X}_2) - (\mu_1 - \mu_2)}{\sqrt{\dfrac{\sigma_1^2}{n_1} + \dfrac{\sigma_2^2}{n_2}}}$$

Normally the hypothesis states that $\mu_1 - \mu_2 = 0$, in which case the term $\mu_1 - \mu_2$ will drop out of the preceding formula. Assumes σ_1 and σ_2 are known.

Test for proportions

$$Z = \frac{X - nP}{\sqrt{nPQ}}$$

Assumes n is reasonably large.

Test for the difference of two proportions

$$Z = \frac{(P_1 - P_2) - (p_1 - p_2)}{\sqrt{\dfrac{P_1 Q_1}{n_1} + \dfrac{P_2 Q_2}{n_2}}}$$

The common hypothesis states that $P_1 - P_2 = 0$, in which case $p_1 - p_2$ drops out of the preceding formula. Assumes n_1 and n_2 are reasonably large.

Test statistic Z, difference in means: known population variances

$$Z = \frac{\bar{X}_1 - \bar{X}_2}{\sqrt{\dfrac{\sigma_1^2}{n_1} + \dfrac{\sigma_2^2}{n_2}}}$$

Test statistic Z, difference in proportion

$$Z = \frac{(P_1 - P_2) - (p_1 - p_2)}{\sqrt{\dfrac{P_2 Q_1}{n_2} + \dfrac{P_2 Q_2}{n_2}}}$$

ORGANIZED LEARNING QUIZ

1. We wish to decide if the average output of Machine New (1) is greater than the average output of Machine Old (2), which is 125 units per day. The appropriate null and alternate hypotheses are

   a. $H_0: \mu_1 = \mu_2$; $H_1: \mu_1 \neq \mu_2$.
   b. $H_0: \mu_2 = 125$; $H_1: \mu_1 > 125$.
   c. $H_0: \mu_1 = \mu_2$; $H_1: \mu_1 > \mu_2$.
   d. none of the above.

2. The probability of making a type I error is

   a. α.
   b. β.
   c. $1 - \alpha$.
   d. unimportant

3. A correctly executed test of a statistical hypothesis establishes the truth of either the null hypothesis or the alternate hypothesis.

   a. True
   b. False

4. For a statistical test to be logically correct, the probability of a type I error must be less than the probability of a type II error.

   a. True
   b. False

5. One can never prove the truth of a hypothesis; one can only tend to discount or support it.

   a. True
   b. False

6. We have defined the power of a test as the probability of rejecting the hypothesis.

   a. True
   b. False

7. A Z score represents the deviation of a specific score from the mean expressed in standard deviation units.

   a. True
   b. False

8. The probability of a type II error is

   a. .01.
   b. .05.
   c. α.
   d. β.
   e. $1 - \alpha$.

9. Given that the null hypothesis is true, the probability of making a type II error is

   a. $1 - \alpha$.
   b. .00.
   c. 1.00.
   d. $1 - \beta$.

10. The critical region is

   a. the confidence interval.
   b. a law stating that the form of the sampling distribution is normal.
   c. that portion of the area under the curve that includes those values of a statistic leading to rejection of H_0.
   d. none of the above.

11. Alpha is the probability that _____; beta is the probability that _____.

12. If H_0: $\mu = 1000$ hours and H_1: $\mu > 1000$ hours, the decision rule would be _____, where C is a sufficiently _____ value for the test statistic $\bar{X}$ to cause the rejection of the (null/alternate) hypothesis.

13. H_1: $\mu > .05$ is a _____-tailed test, whereas

    H_1: $\mu \neq .05$ is a _____-tailed test.

14. Alpha is involved only when _____: beta is involved only when _____.

15. H_0: $\mu = .05$ is a(n) _____ hypothesis;

    H_0: $\mu > .05$ is a(n) _____ hypothesis.

16. The output of workers in Department A is known to be normally distributed with a mean of 625 units per week and a variance of 121. A sample of 25 employees who have just completed a training program is selected. A decision rule to test the hypothesis that the mean output of employees who have taken the training program is significantly greater than 625 at $\alpha = .05$ would be _____.

17. On the basis of the decision rule in problem 16, would you reject the null hypothesis if the sample yields a mean of 630? _____

18. A manufacturing process is designed to bore a hole 1.546 inches from the edge of a piece of metal. The variance in the process, measured by the standard deviation is .008 inch. A sample of four pieces taken at random from this assembly line is found to have a mean of 1.566 inches. Is the difference between the sample mean and the universe value of 1.546 inches significant? _____

19. Two lots of cotton yarn were purchased from different mills. It was desired to test whether there was a significant difference in the strength of the two lots. A sample of ten

skeins from each lot was tested with the result that the mean of lot 1 was 95 pounds of pressure required to break the yarn; for lot 2, the mean was 82.4 pounds. Assume that it is known that the variance for the two mills' manufacturing processes is $\sigma_1^2 = 6.2$ and $\sigma_2^2 = 2.4$. Is the difference significant at the .05 level? _____

20. Barry Bizarre of Scary Masks, Inc., wishes to determine if his new Halloween mask is sufficiently scary to be included in his product line. If it impresses more than 50% of the children who see it, he will add it to his line. (Use a significance level of .05.) Barry's sales manager tested the mask on 20 children and concluded that 14 were impressed. Should Barry add the mask to his product line? _____

21. List the steps in hypothesis testing.

22. List the four possible decision rule results from a hypothesis testing problem.

23. Explain the process of treating an inexact null hypothesis as an exact null hypothesis.

24. What is the distinction between the parametric and nonparametric approaches to statistical testing?

25. Identify the two approaches to testing hypotheses dealing with sample proportions.

ANSWERS TO ORGANIZED LEARNING QUIZ

1. c.
2. a.
3. b.
4. b.
5. a.
6. b.
7. a.
8. d.
9. b.
10. c.

11. H_0 is rejected when it is true; H_0 is not rejected when it is false.

12. Reject H_0 if $\bar{X}$ is $\geq C$, large, null.

13. one, two

14. The null hypothesis is true; the alternative hypothesis is true.

15. exact; inexact

16. Reject H_0 if $\dfrac{\bar{X} - 625}{\sqrt{121/25}} \geq 1.96$.

17. Yes, $\bar{X} = 630$; $\dfrac{630 - 625}{\sqrt{121/25}} = \dfrac{5}{2.2} = 2.27$.

18. Yes, $Z = \dfrac{1.566 - 1.546}{.008/\sqrt{4}} = \dfrac{.020}{.004} = 5$.

    No alpha given; assuming $\alpha = .01$, the decision rule becomes:

    Reject H_0 if $\dfrac{\bar{X} - 1.546}{.008/\sqrt{4}} \geq 2.58$.

    Therefore reject the null hypothesis.

19. Yes, $Z = \dfrac{\bar{X}_1 - \bar{X}_2}{\sqrt{\dfrac{\sigma_1^2}{n_1} + \dfrac{\sigma_2^2}{n_2}}} = \dfrac{95.0 - 82.4}{\sqrt{\dfrac{6.2}{10} + \dfrac{2.4}{10}}} = \dfrac{12.6}{.927} = 13.6$

    When $\alpha = .05$, the decision rule is

    Reject H_0: $\mu_1 - \mu_2 = 0$ if $Z \geq 1.96$ or $Z \leq 1.96$.

20. a. The hypotheses involved one:

    H_0: $p \leq 0.5$ (Treat as $p = .5$.)

    H_1: $p > 0.5$

    b. critical value of $\bar{X} = np + 1.96\sqrt{npq}$
    $= 10 + 1.96\sqrt{5} = 14.3$

    A value of $\bar{X} = 14 < 14.3$ fails to reject the null hypothesis that it impresses no more than 50% of children.

test statistic:

$$Z = \frac{X - np}{\sqrt{npq}} = \frac{14 - (20)(.5)}{\sqrt{20(.5)(.5)}} = \frac{10 - 10}{2.24} = 1.79 < 1.96$$

cannot reject H_0 and should <u>not</u> add to collection.

21.
    1. Identification of the pattern of the population distribution.
    2. Formulation of the null and alternative hypotheses.
    3. Specification of the level of significance.
    4. Formulation of the decision rule.
    5. Decision making: the task is finished when you decide whether the null hypothesis is rejected.

22.

Decision	H_0 True	H_0 False
Reject H_0	Type I error (α)	Correct ($1 - \beta$)
Do not reject H_0	Correct ($1 - \alpha$)	Type II error (β)

23. If the hypothesis is in the form H_1: $\mu_1 \neq \mu_2$, they are inexact. The hypotheses can be treated as exact and the comparison made between the Z value and the critical value.

24. Using the parametric approach, assumptions are made about the nature of the underlying population distribution. With the nonparametric approach, no specification of the population distribution is made.

25. The difference is sample size. For small values of n, use the cumulative binomial probability table in Appendix C. For large values, use the method of normal approximation.

SAMPLE EXERCISE

Refer to the Sample Exercise in Chapter 9. Once Ms. Precise has collected her data, she may want to test the hypothesis that the mean error rates for the two types of forms are the same. Let μ_1 be the mean error rate for the old form, μ_2 the mean for the new form.

Step 1. Assume population is normally distributed.

Step 2. H_0: $\mu_2 - \mu_1 = 0$

H_1: $\mu_2 - \mu_1 \neq 0$

Step 3. Set $\alpha = .05$.

Step 4. Reject H_0 if $Z \geq 1.96$ or $Z \leq -1.96$.

Step 5. Decision

$$Z = \frac{\bar{X}_1 - \bar{X}_2}{\sqrt{\frac{\sigma_1^2}{n_1} + \frac{\sigma_2^2}{n_2}}} = \frac{.020 - .012}{\sqrt{\frac{.005}{16} + \frac{.010}{16}}} = \frac{.008}{.031} = .258$$

Therefore do not reject the null hypothesis.

CASE PROBLEM

Refer back to the case presented at the end of Chapter 9. Using the five steps in hypothesis testing, help the sales manager decide if the average sales performance of the two salesmen is the same.

11

STUDENT'S t TESTS

CHAPTER LEARNING OBJECTIVES

On completing this chapter, a student should be able to

- Describe the characteristics of the t curve.

- List the similarities and differences between the t distribution and the z distribution.

- Explain the difference in the procedure for testing hypotheses about population means for small and large samples.

- List the assumptions of two sample tests for population means when the samples are independent.

- Explain the conditions under which the Z statistic can be used as an approximation to the T statistic.

- Estimate confidence intervals for population means and differences when the population variance is not known.

SYNOPSIS

The t Distribution

The ratio used for the Z statistic involves only one random variable along with two parameters. It is based on a known population standard deviation. The table of areas under the standard normal curve, however, is theoretically correct if the true value of σ is known. If only a sample standard deviation is available, the ratio follows a t distribution with n - 1 degrees of freedom.

The t distribution is a family of symmetrical curves much like the shape of a normal curve but flatter and approaches the normal curve as the sample size, and hence the degrees of freedom, increases. Since the t distribution is a family of curves depending on n, most t tables show selected critical values with their selection dependent on a specific level of significance and degree of freedom. The critical values of t cut off the rejection region from the rest of the area and can be used in one-tail and two-tail tests.

As will be shown later in the chapter, for large sample sizes, 30 or larger, the standard normal distribution can be used as an approximation to the t distribution for hypothesis testing purposes.

The Population Mean

Small samples are likely to have an S^2 a great deal of difference from σ^2 so that inferences using Z scores could contain serious errors. In these cases, the T statistic, where the only assumption necessary is that the population be normally distributed, should be used.

Testing for the Difference Between Two Means

The question may arise as to whether the means of two separate normal populations are significantly different and at least one of the two sample sizes taken from the populations is small. The t test is appropriate in this case.

Independent Samples. Care must be taken to determine whether the two populations are statistically dependent or independent. If independent, then the assumption of equal variances must also be assumed, for then the variance of the difference between the two sample means is the sum of the variances of the means.

$$\sigma_{\bar{D}}^2 = \frac{\sigma_1^2}{n_1} + \frac{\sigma_2^2}{n_2} = \sigma^2\left(\frac{1}{n_1} + \frac{1}{n_2}\right)$$

By replacing σ^2 by its best estimator S^2, we can estimate the variance of the difference between two sample means, $S_{\bar{D}}^2$.

We pool information from the samples to get a better estimation of σ^2. The pooled estimator S^2 is obtained by summing the two sample variations (sums of squares) and dividing by the pooled degrees of freedom.

$$S_1^2 = \frac{\Sigma (X_1 - \bar{X}_1)^2}{n_1 - 1} = \frac{\text{variation of sample 1}}{\text{df of sample 1}}$$

$$S_2^2 = \frac{\Sigma (X_2 - \bar{X}_2)^2}{n_2 - 1} = \frac{\text{variation of sample 2}}{\text{df of sample 2}}$$

$$S^2 = \frac{\Sigma (X_1 - \bar{X}_1)^2 + \Sigma (X_2 - \bar{X}_2)^2}{(n_1 - 1) + (n_2 - 1)} = \frac{S_1^2(n_1 - 1) + S_2^2(n_2 - 1)}{n_1 + n_2 - 2}$$

The square root of $S_{\bar{D}}^2$ is $S_{\bar{D}}$, the standard error of the difference between two sample means. This is the quantity used in the T ratio when testing hypotheses about the difference between two population means.

$$T = \frac{\bar{X}_1 - \bar{X}_2}{S_{\bar{D}}} = \frac{\bar{X}_1 - \bar{X}_2}{\sqrt{S^2\left(\frac{1}{n_1} + \frac{1}{n_2}\right)}} = \frac{\bar{X}_1 - \bar{X}_2}{\sqrt{\frac{S_1^2(n_1 - 1) + S_2^2(n_2 - 1)}{n_1 + n_2 - 2}\left(\frac{1}{n_1} + \frac{1}{n_2}\right)}}$$

The test statistic is then compared with the critical t value for the appropriate level of significance and $n_1 + n_2 - 2$ degrees of freedom.

Notice that when $n_1 = n_2 = n$, then the T ratio becomes

$$T = \frac{\bar{X}_1 - \bar{X}_2}{\sqrt{\frac{S_1^2(n-1) + S_2^2(n-1)}{2(n-1)}\left(\frac{2}{n}\right)}} = \frac{\bar{X}_1 - \bar{X}_2}{\sqrt{\frac{S_1^2 + S_2^2}{n}}} .$$

Dependent Samples. If samples are taken pairwise and the second observation is someway related to or dependent on the selection of the first, then the two observations are said to be dependent. Knowing the value of the first helps to predict the value of the second.

In this case, the variance of the difference between sample means is no longer obtained as in the preceding section. If D_i is the difference between the two observations in the matched pair i, then

$$S_{\bar{D}}^2 = \frac{S_D^2}{n-1} = \frac{\text{variation of D}}{n-1} = \frac{\Sigma D^2 - \bar{D}\Sigma D}{n(n-1)}$$

$$S_{\bar{D}} = \sqrt{S_{\bar{D}}^2} .$$

T is calculated as

$$T = \frac{\bar{D}}{S_{\bar{D}}} .$$

KEY TERMS

Confidence interval   The interval whereby the true value of the parameter may lie with a probability equal to one minus the level of significance.

Degrees of freedom	The sole parameter of the t distribution. Equal to (n - 1).
Dependent samples	Two samples where the observations in one are paired with and related to a corresponding observation in the other.
Hypothesis	An assumption about the form of a population or its parameters.
Independent samples	Two samples where the observations in one are not related to the observations in the other.
Level of significance	The probability of rejecting a true hypothesis. This probability of error is determined by the reasearcher preferably before the experiment.
Nonparametric methods	Statistical methods that are not necessarily based on the assumption of a normal population distribution.
Rejection region	The set of outcomes for the test statistic that leads to the rejection of the hypothesis.
t critical value	The table value for t that separates the rejection region of the appropriate t distribution from the nonrejection region.
t distribution	A sampling distribution of means where the population parameter for the standard deviation is replaced by a sample standard deviation. The correct distribution to use no matter what the sample size; however, for large sample sizes (over 30), the standard normal distribution becomes a good approximation.
T statistic	A statistic used in testing hypotheses involving the t distribution.
Test	A rule or procedure used for deciding whether to accept or reject a hypothesis.
Test statistic	A quantity computed from a sample used in conjunction with a test and a rejection region to decide on the acceptance or rejection of a hypothesis.
Variation	The sums of squares of the deviations of a variable from its mean. When averaged, the variation becomes a variance.

FORMULA REVIEW

(11-1)   T test statistic $T = \dfrac{\bar{X} - \mu}{S/\sqrt{n}}$

(11-2) Confidence interval for μ:
$$\bar{X} - t_{(\nu,\alpha/2)} \frac{S}{\sqrt{n}} < \mu < \bar{X} + t_{(\nu,\alpha/2)} \frac{S}{\sqrt{n}}$$

(11-13)
$$\bar{X} - Z_{\alpha/2} \frac{S}{\sqrt{n}} < \mu < \bar{X} + Z_{\alpha/2} \frac{S}{\sqrt{n}}$$

(11-5) T ratio for testing the hypothesis about the difference between two population means.

(independent samples, unequal sample sizes)

$$T = \frac{\bar{X}_1 - \bar{X}_2}{\sqrt{\frac{(n_1-1)S_1^2 + (n_2-1)S_2^2}{n_1 + n_2 - 2} \left(\frac{1}{n_1} + \frac{1}{n_2}\right)}}$$

(independent samples, equal sample sizes)

$$T = \frac{\bar{X}_1 - \bar{X}_2}{\sqrt{\frac{S_1^2 + S_2^2}{n}}}$$

(dependent samples)

$$T = \frac{\bar{D}}{\sqrt{\frac{\Sigma D^2 - \bar{D} \Sigma D}{n(n-1)}}}$$

(11-6) Confidence interval for δ (the true difference between two populations means)

$$\bar{D} - t_{(\nu,\alpha/2)} S_{\bar{D}} < \delta < \bar{D} + t_{(\nu,\alpha/2)} S_{\bar{D}}$$

$$\bar{D} - Z_{\alpha/2} \sqrt{\frac{S_1^2}{n_1} + \frac{S_2^2}{n_2}} < \delta < \bar{D} + Z_{\alpha/2} \sqrt{\frac{S_1^2}{n_1} + \frac{S_2^2}{n_2}}$$

(11-4) Standard error of difference between two sample means

(independent samples, unequal sample sizes)

$$S_{\bar{D}} = \sqrt{\frac{S_1^2(n_1-1) + S_2^2(n_2-1)}{n_1 + n_2 - 2} \left(\frac{1}{n_1} + \frac{1}{n_2}\right)}$$

(independent samples, equal sample sizes)

$$S_{\bar{D}} = \sqrt{\frac{S_1^2 + S_2^2}{n}}$$

(dependent samples)

$$S_{\bar{D}} = \sqrt{\frac{\Sigma D^2 - \bar{D}\, \Sigma D}{n(n-1)}}$$

Normal Approximation--sample size large (more than 30)
Tests about the population mean

$$Z = \frac{\bar{X} - \mu}{S/\sqrt{n}}$$

Tests about the difference between two means

$$Z = \frac{\bar{X}_1 - \bar{X}_2}{\sqrt{\frac{S_1^2}{n_1} + \frac{S_2^2}{n_2}}} \qquad \text{(independent samples)}$$

ORGANIZED LEARNING QUIZ

1. As the sample size increases when sampling from a normal population distribution,

    a. the population changes.
    b. the z distribution changes.
    c. the t distribution changes.
    d. the population parameters change.

2. When sampling from a normal population, the T statistic

    a. is only valid for small sample sizes.
    b. is not as valid as the Z statistic for large sample sizes.
    c. is not as valid as the Z statistic for small sample sizes.
    d. is preferred to the Z statistic for small sample sizes.

3. When testing hypotheses concerning a population mean,

    a. one must use the t test at all times.
    b. one should use the t test when the population variance is unknown.
    c. one should use the t test when the sample size is over 30.
    d. one should use the t test when the population is not normally distributed.

4. For independent samples, the variance of the difference between two sample means

   a. is the difference of the means of the samples.
   b. is the sum of the means of the samples.
   c. is the difference of the variances of the sample means.
   d. is the sum of the variances of the sample means.

5. For a given level of significance,

   a. the critical values of t are greater than z for all sample sizes.
   b. the critical values of t are less than z for all sample sizes.
   c. the t values are equal to the z values for all sample sizes.
   d. sometimes the critical t values are larger and sometimes smaller, depending on the sample size.

6. One-tail tests

   a. are never appropriate when using the T statistic.
   b. can be used with the T statistic, but results are unreliable.
   c. are appropriate for some T-statistic hypothesis testing.
   d. are used for all T-statistic hypothesis testing.

7. When the sample size is sufficiently large and the population variance is unknown,

   a. the z test can be used chiefly because it is a good approximation to the more accurate test
   b. the z test is always better to use because the t test becomes inaccurate for large sample sizes.
   c. neither the z test nor the t test should be used.
   d. z tests cannot be used, since z values are not available for large sample sizes and are never appropriate when the population variance is not known.

8. For a given sample size,

   a. there is a different t distribution for each level of significance.
   b. all sample means are constant.
   c. there is a single t distribution.
   d. the standard deviation is $\sqrt{n}$.

9. The standard error of the difference refers to

   a. the standard deviation of the differences between two population means.
   b. the standard deviation of the differences between two sample observations.
   c. the standard deviation of the differences between two sample variances.
   d. the standard deviation of the differences between two sample means.

10. Critical t values

    a. are larger for larger levels of significance.
    b. are larger for smaller levels of significance.
    c. are not dependent on levels of significance.
    d. are always one-half the critical z values.

11. As the sample size is increased, the critical t values for any specific level of significance _____ in value until when the sample size is infinite the critical t value becomes the same as _____.

12. In order to determine which particular t distribution to use in a particular test, one only needs to know the _____.

13. In a two-tailed test the corresponding critical t value is _____ than that of a one-tail test. (larger, smaller)

14. When testing hypotheses about the differences between two population means, while using independent samples, the two assumptions are _____ and _____.

15. If it is desirable to compare the scores of individuals before and after some training, then the two samples of scores (before and after) are said to be _____.

16. Explain what is meant by a pooled variance.

    _____
    _____
    _____
    _____

17. Explain why the Z statistic can be used as an approximation to the T statistic in some cases.

    _____
    _____
    _____

18. Explain why the T statistic can still be used for two sample tests even though the samples are dependent.

    _____
    _____
    _____

19. Explain why the Z statistic is not appropriate when the population variance is not known and the sample size is small.

    _____

20. Can the two sample tests explained in this chapter be used on a single population rather than on two populations? Explain.

ANSWERS TO ORGANIZED LEARNING QUIZ

1. c
2. d
3. b
4. d
5. a
6. c
7. a
8. c
9. d
10. b

11. decreases, the critical z value

12. sample size (degrees of freedom)

13. larger

14. normal populations and equal variances

15. dependent

16. Since the assumption is made that the variances of the two populations are equal and the samples were derived independently of each other, the two samples can be combined (pooled). They are then used as a single larger sample to better estimate what the population variance is. A pooled variance is then a variance of the combination of the two samples.

17. When hypothesis testing under situations where the population variance is unknown, the T statistic should be used. However, as the sample size increases, the critical t values approach the z value so that, for large sample sizes, the z value becomes a good approximator.

18. The T ratio included a measure of the standard error of the differences in the denominator. It is this measure that is calculated differently for independent and dependent samples.

19. When the sample size is large, deviations of the sample means from their true mean are approximately normal and the Z statistic may be used. When the population variance is not known, then the standard error of the sample means may be satisfactorily approximated, using the sample variance. However, for

small samples, the ratio used to calculate Z does not follow the values of the Z distribution but takes on values approaching it as the sample size increases.

20. Testing differences between two samples from the same population receiving two different treatments or the same elements before and after some treatment still results in making inferences about two separate populations.

SAMPLE EXERCISE

A large fast-food chain is faced with the training of a group of 150 prospective store managers. In the past the training has been done on the job with assignments as assistant managers in actual working conditions. The training supervisor has made a proposal to management to bring the trainees to the home office and enroll them in a company-run intensive program with the belief that they will be more successful in the training process.

A random sample of 16 trainees is carefully paired in order to eliminate the influences other factors may have on the objective score achieved by the trainees at the end of their training. These factors might be age, education, previous experience, or other background information. One member of each pair is sent to the company school while the other is assigned a position as an assistant manager on the job.

The experimental hypothesis in this case is whether the mean scores of the two groups are significantly different.

The two samples are certainly dependent, since they were intentionally paired to eliminate causes that might explain why pairs of trainees should receive wide variations in scores.

The hypotheses are

$$H_o: \mu_1 = \mu_2$$

$$H_1: \mu_1 \neq \mu_2$$

Hence the investigator would use a two-tailed test. The following table lists the data and the calculations follow.

Scores of A	Scores of B	D = Difference	D-square
112	105	7	49
86	94	-8	64
94	87	7	49
78	73	5	25
103	91	12	144
88	81	7	49
64	73	-9	81
97	83	14	196
Total		35	657
Means and corrections		4.375	151.125

127

$$S_D^2 = 503.875/7 = 71.98$$

$$S_D = 8.484 \quad S_{\bar{D}} = 3$$

$$T = 4.375/3 = 1.458$$

The table value $t_{(7, .05)} = 2.365$; therefore fail to reject.

CASE PROBLEM

Lyons Manufacturing

The management of Lyons Manufacturing has definitly made the decision to give the go-ahead on implementing a new line of product in direct competition with its leading competitor. The R and D department has come forth with prototypes that are in the same price range as the competitor's, but the quality as measured against their product is questionable. Roger Goslin, a recent graduate, has been given the assignment of determining whether the company's new product will perform favorably with the competitor's product.

After studying the requirements of management and determining the importance that the results of his test will have on future decisions, Roger decides to use a significance level of the null hypothesis to be tested as "There is no statistically significant difference between the average performance of our product A and the competitor's product X."

Lyon's engineering group has designed a test whereby one year's normal usage of the products can be simulated in a matter of days. One hundred of the competitor's product are randomly obtained and subjected to the simulated performance test. The results are summarized in Exhibit A.

Exhibit A
Summary of Simulated Performance Results

	Product X	Product A
Size of sample	100	400
Mean of sample	325	320
Standard deviation	40	35

Roger was apprehensive about the favorable mean performance of the random sample of product X but was pleased that product A was slightly more consistant in its measurement. The question he had to answer: Can the difference in sample means be due to chance or is there a real difference in population means?

The two random samples were taken independently and the assumptions of normal populations with equal variances were made. Roger wondered about the assumption of equal variances of the two populations when there was some difference in the two sample variances.

He reasoned, however, that if he were to concede that the two populations were equal and could therefore be considered as one population, rejection of the null hypothesis of equal means would be more difficult and would carry more credibility.

The null hypothesis of no difference in means implies that sometimes the difference of means would be positive and sometimes negative but that over many samples the average difference would be zero. If those differences of means could be tabulated for many samples, their standard deviation could be computed and used to establish critical regions for hypothesis testing. However, Roger has only the two samples and thus he will have to estimate the standard deviation of the mean differences with his sample data. The formula he uses is based on the use of pooled information about the population variance and the ratio for the z test. In this situation, the Z statistic, instead of

$$Z = \frac{\text{statistic} - \text{parameter}}{\text{standard deviation of the statistic}},$$

becomes

$$Z = \frac{(\text{observed difference}) - (\text{hypothesized difference})}{\text{standard error of the difference}}$$

and the standard error of the difference is

$$S_{\overline{D}} = \sqrt{\frac{S_1^2}{n_1} + \frac{S_2^2}{n_2}}.$$

Roger's calculations are

$$Z = \frac{325 - 320}{\sqrt{\frac{40^2}{100} + \frac{35^2}{400}}} = 1.147$$

$$Z_{.025} = 1.96.$$

Therefore fail to reject; there is no significant difference at the 5% level.

Because of random causes, the results obtained with the experiment could indeed occur and Roger cannot assume the difference in the sample means was because of a difference in the products.

COMPUTER PROGRAM

On the following pages a flowchart, BASIC program listing, and results of the Sample Exercise are given. The program allows the user to select either a one-sample test or a two-sample test; if two samples, then either dependent or independent samples are allowed. If any of the samples entered have less than 30 observations, then the output is labeled a T statistic. Over 30 observations generates a Z statistic. The dimension statements allow a maximum sample size of 50. These can be altered by the user.

Flowchart for Computer Program

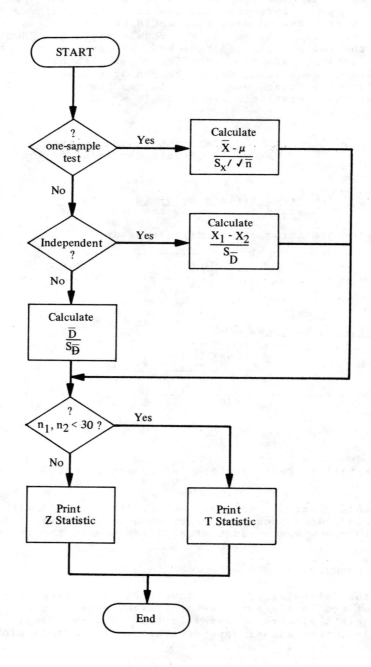

```
10 REM****                 ****************************************
20 REM
30 REM       TITLE:       MEANTEST
40 REM       LANGAUGE:    BASIC
50 REM       DESCRIPTIONS:
60 REM          THIS PROGRAM ALLOWS THE USER TO SELECT
70 REM          EITHER ONE-SAMPLE TEST OR TWO-SAMPLE
80 REM          ; IF TWO SAMPLES, THEN EITHER DEPENDENT
90 REM          OR INDEPENDENT SAMPLES ARE ALLOWED.
100 REM         IF ANY OF THE SAMPLES ENTERED HAVE
110 REM         LESS THAN 30 OBSERVATIONS, THEN THE
120 REM         OUTPUT IS LABELED A T-STATISTIC.
130 REM         OVER 30 OBSERVATIONS GENERATES A
140 REM         Z STATISTIC.
150 REM      INSTRUCTIONS:
160 REM         ENTER DATA AS REQUESTED BY THE PROGRAM.
170 REM      PROGRAMMER:
180 REM         GROVER WM. RODICH, PORTLAND STATE UNIV.
190 REM
200 REM*************************************************************
210 REM
220 REM
230 KEY OFF
240 DIM X(50),Y(50)
250 DIM D(50),D2(50)
260 CLS
270 PRINT "THIS PROGRAM CALCULATES EITHER THE T STATISTIC OR THE "
280 PRINT "Z STATISTIC TO USE WHEN TESTING HYPOTHESES ABOUT MEANS"
290 PRINT "TYPE 1 AND RETURN FOR A ONE SAMPLE TEST OR 2 AND RETURN"
300 PRINT "FOR A TWO SAMPLE TEST.  (TYPE O AND RETURN TO EXIT)"
310 INPUT A
320 IF A=0 THEN 1380
330 IF A=2 THEN 670
340 IF A<>1 THEN 290
350 CLS
360 PRINT "FOR A ONE SAMPLE TEST YOU ARE TESTING WHETHER THE POPULATION"
370 PRINT "MEAN IS DIFFERENT FROM WHAT HYPOTHESIZED VALUE";
380 INPUT A
390 PRINT "WHAT IS THE SAMPLE SIZE";
400 INPUT N
410 PRINT "INPUT EACH DATA ITEM ONE AT A TIME"
420 PRINT "WHAT IS THE FIRST ITEM";
430 INPUT X(1)
440 PRINT "CONTINUE--SECOND ITEM";
450 INPUT X(2)
460 FOR I=3 TO N
470 INPUT X(I)
480 NEXT I
490 PRINT "THANK YOU, I AM NOW COMPUTING"
500 T=0
510 T2=0
520 FOR I=1 TO N
530 T=T+X(I)
540 T2=T2+X(I)*X(I)
550 NEXT I
560 M=T/N
570 V=T2-M*T
580 V2=V/(N-1)
590 S=SQR(V/(N-1))
600 IF N<30 THEN 640
610 Z=(M-A)/(S/N^.5)
620 PRINT "THE CALCULATED Z STATISTIC IS";Z
630 GOTO 1360
640 T=(M-A)/(S/N^.5)
650 PRINT "THE CALCULATED T STATISTIC IS";T
660 GOTO 1360
670 CLS
680 PRINT "INPUT 1 AND RETURN FOR DEPENDENT SAMPLES OR 2 AND RETURN"
690 PRINT "FOR INDEPENDENT SAMPLES"
700 INPUT A
710 IF A=2 THEN 730
720 IF A<>1 THEN 680
730 PRINT "INPUT SIZES OF SAMPLES SEPARATED BY A COMMA";
```

```
740  INPUT N1,N2
750  PRINT "INPUT FIRST SAMPLE ONE AT A TIME"
760  PRINT "FIRST ITEM";
770  INPUT X(1)
780  PRINT "NEXT";
790  INPUT X(2)
800  FOR I=3 TO N1
810  INPUT X(I)
820  NEXT I
830  PRINT "INPUT SECOND SAMPLE ONE AT A TIME"
840  PRINT "FIRST ITEM";
850  INPUT Y(1)
860  PRINT "NEXT";
870  INPUT Y(2)
880  FOR I=3 TO N2
890  INPUT Y(I)
900  NEXT I
910  IF A=2 THEN 1100
920  T=0
930  T2=0
940  FOR I=1 TO N1
950  D(I)=X(I)-Y(I)
960  T=T+D(I)
970  D2(I)=D(I)*D(I)
980  T2=T2+D2(I)
990  NEXT I
1000 M=T/N1
1010 V=T2-(M*T)
1020 V2=V/(N1-1)
1030 S=V2^.5
1040 G=M/(S/N1^.5)
1050 IF N1>30 THEN 1080
1060 PRINT "THE T STATISTIC IS";G
1070 GOTO 1360
1080 PRINT "THE Z STATISTIC IS";G
1090 GOTO 1360
1100 T=0
1110 T2=0
1120 FOR I=1 TO N1
1130 T=T+X(I)
1140 T2=T2+(X(I)*X(I))
1150 NEXT I
1160 M1=T/N1
1170 K=T2
1180 H1=T/N1
1190 T=0
1200 T2=0
1210 FOR I=1 TO N2
1220 T=T+Y(I)
1230 T2=T2+Y(I)*Y(I)
1240 NEXT I
1250 M2=T/N2
1260 H2=(T/N2)*T
1270 S=(K+T2-H1-H2)/(N1+N2-2)
1280 S=S^.5
1290 S=S*((1/N1+1/N2)^.5)
1300 G=(M1-M2)/S
1310 IF N1>30 THEN 1340
1320 PRINT "THE T STATISTIC IS";G
1330 GOTO 1360
1340 IF N2<30 THEN 1320
1350 PRINT "THE Z STATISTIC IS";G
1360 LOCATE 24,1:PRINT "TO CONTINUE PRESS ANY KEY";:A$=INPUT$(1)
1370 GOTO 260
1380 KEY ON:END
```

```
THIS PROGRAM CALCULATES EITHER THE T STATISTIC OR THE
Z STATISTIC TO USE WHEN TESTING HYPOTHESES ABOUT MEANS
TYPE 1 AND RETURN FOR A ONE SAMPLE TEST OR 2 AND RETURN
FOR A TWO SAMPLE TEST.  (TYPE O AND RETURN TO EXIT)
? 2


INPUT 1 AND RETURN FOR DEPENDENT SAMPLES OR 2 AND RETURN
FOR INDEPENDENT SAMPLES
? 1
INPUT SIZES OF SAMPLES SEPARATED BY A COMMA? 8,8
INPUT FIRST SAMPLE ONE AT A TIME
FIRST ITEM? 112
NEXT? 86
? 94
? 78
? 103
? 88
? 64
? 97
INPUT SECOND SAMPLE ONE AT A TIME
FIRST ITEM? 105
NEXT? 94
? 87
? 73
? 91
? 81
? 73
? 83
THE T STATISTIC IS 1.458514
TO CONTINUE PRESS ANY KEY


THIS PROGRAM CALCULATES EITHER THE T STATISTIC OR THE
Z STATISTIC TO USE WHEN TESTING HYPOTHESES ABOUT MEANS
TYPE 1 AND RETURN FOR A ONE SAMPLE TEST OR 2 AND RETURN
FOR A TWO SAMPLE TEST.  (TYPE O AND RETURN TO EXIT)
? 0
OK
```

12

CHI-SQUARE TESTS

CHAPTER LEARNING OBJECTIVES

On the completion of this chapter the student should be able to

- Explain the assumptions necessary when using a chi-square test.

- Describe the characteristic of the χ^2 distribution.

- List the three major tests where a chi-square distribution is used.

- List the three common null hypotheses used when testing for attributes of a population variance.

- Construct and explain the parts of a contingency table.

- Explain the concept of statistical independence.

SYNOPSIS

χ^2 *Distribution*

As with the t distribution, χ^2 (chi-square) has one parameter (degrees of freedom) and is therefore a whole family of curves. Unlike the symmetrical t distribution, the χ^2 distribution is positively skewed. As the degrees of freedom increases, the χ^2 distribution becomes more symmetrical, approaching the shape of a normal curve. The chi-square curve is nonnegative, however, and so theoretically it cannot exactly duplicate the normal curve.

Tables of critical values for commonly used significant levels and degrees of freedom are used similar to the tables for critical values of t. Care must be taken when using an unfamiliar table to ensure that its interpretation is correct.

If a random sample is taken from a normal population with

variance σ^2, then it can be mathematically shown that

$$X^2 = \frac{(n-1)s^2}{\sigma^2}$$

is distributed as χ^2 with $n - 1$ degrees of freedom.

Testing For a Single Variance

On closer inspection of the X^2 ratio, we can see that s^2 only differs from X^2 by the constant $(n-1)/\sigma^2$. Thus the sampling distribution of s^2 is closely related to the χ^2 distribution of X^2. Consequently, the χ^2 distribution can be used to test hypotheses about a population variance on the basis of the value of s^2.

The null hypothesis and the alternate can usually be written in one of three ways:

(1) $H_0: \sigma^2 = \sigma_0^2$ (2) $H_0: \sigma^2 \geq \sigma_0^2$ (3) $H_0: \sigma^2 \leq \sigma_0^2$

$H_1: \sigma^2 \neq \sigma_0^2$ $H_1: \sigma^2 < \sigma_0^2$ $H: \sigma^2 > \sigma_0^2$

High and low values of X^2 would lie in the critical regions in case (1). In case (2), only small values of X^2 contradict H_0 and in case (3) only high values of X^2 contradict H_0. The following table summarizes these statements.

Case (1)	Case (2)	Case (3)
$X^2 < \chi^2_{n-1,\alpha/2}$ or $X^2 > \chi^2_{n-1,1-\alpha/2}$	$X^2 < \chi^2_{n-1,\alpha}$	$X^2 > \chi^2_{n-1,1-\alpha/2}$

The $1 - \alpha$ confidence interval for the parameter σ^2 based on s^2 is

$$\frac{(n-1)s^2}{\chi^2_{n-1,\alpha/2}} < \sigma^2 < \frac{(n-1)s^2}{\chi^2_{n-1,1-\alpha/2}}.$$

Testing for Goodness of Fit

In the goodness-of-fit test inferences are made about an **entire** population distribution rather than the attributes of the population parameters. By taking a sample and using the test, a decision can be made as to whether the sample distribution is the one specified in the null hypothesis. Sample data need to be classified into an observed frequency distribution, which is then compared to an expected frequency distribution. If the n observations are classified into J classes, then

$$X^2 = \Sigma \frac{(O_j - E_j)^2}{E_j}$$

where O stands for observed frequencies and E stands for expected frequencies. X^2 is distributed as chi-square with J - 1 degrees of freedom. Care should be taken that no classification has less than five observations.

Since the summation is small when the fit is good, rejection of the null hypothesis of no differences will occur for abnormally high values of X^2 and the test is a one-tail test.

Testing For Independence

The chi-square procedure can be extended to test whether there is a statistical relationship between two sets of attributes in a population. The first step in the test procedure is to set up a contingency table in which the data are properly classified. After the data are entered in the cross-classified table, the totals of the rows are entered at the right and the totals of the columns are entered at the bottom.

If there were J classifications of one attribute and K classifications of the other, then the table will have J rows and K columns (J x K). The hypothetical data are generated by the assumption of independence. The probability of any item being in a particular row and particular column *at the same time* is, by independence, the product of the probabilities of being in the row and column. The row and column probabilities are calculated by dividing the row and column totals by the grand total. For any cell of the table, the expected entry under independence is then the grand total multiplied by its calculated probability.

The differences between the observed and expected frequencies are next obtained and squared, the squares divided by the expected frequency, and the quotients summed to obtain the calculated X^2 statistic. The number of degrees of freedom is (J - 1) times (K - 1). The previous sentence in formula form becomes

$$X^2 = \sum_{j=1}^{J} \sum_{k=1}^{K} \left[\frac{(O_{jk} - E_{jk})^2}{E_{jk}} \right]$$

If the calculated X^2 is larger than $\chi^2_{(J-1)(K-1),\alpha}$, then the null hypothesis of independent attributes is rejected.

KEY TERMS

Chi-square test   A test to solve problems by comparison of computed chi-square values with theoretical chi-square values. The test is generally discussed under three headings: (1) tests involving one variance, (2) goodness of fit, and (3) independence of principles of classification.

Contingency table   A two-dimensional table of size J by K. Each cell jk contains the frequency of

	observations associated with the jth classification of one attribute of a population and the kth classification of the second attribute. Marginal totals and a grand total are also given.
χ^2 Distribution	A family of continuous, unimodal curves whose probabilities are specified by an equation and one parameter, called degrees of freedom (df). A chi-square variable can take any value from zero to positive infinity; thus the distribution is always skewed to the right.
Expected frequencies	The number of occurrences that would be expected in each classification of an attribute of a population if the population was distributed in some specified theoretical pattern.
Goodness of fit	A test used to determine if observed frequencies of occurrence closely agree with the frequencies that would occur if the distribution followed some theoretical pattern.
Observed frequencies	The observed number of occurrences of a random variable in each classification of an attribute of a population.
Test of statistical independence	A test used to determine whether one variable is independent of another. The chi-square test is used to make a decision about whether some relationship exists between two sets of attributes in a population.

FORMULA REVIEW

(12-1) Chi-square statistic tests involving one variance distributed as χ^2 with $n - 1$ degrees of freedom

$$X^2 = \frac{(n-1)S^2}{\sigma_0^2}$$

X^2 can be calculated as

$$X^2 = \frac{\sum_{i=1}^{n}(X_i - \bar{X})^2}{\sigma_0^2} = \frac{\sum X_i^2 - \bar{X}\sum X_i}{\sigma_0^2}$$

(12-2) Confidence interval for σ^2

$$\frac{(n-1)S^2}{\chi^2_{\nu,\alpha/2}} < \sigma^2 < \frac{(n-1)S^2}{\chi^2_{\nu,1-\alpha/2}}$$

(confidence level of $1 - \alpha$)

(12-3) Test statistic for goodness-of-fit

$$X^2 = \sum_{j=1}^{J} \frac{(O_j - E_j)^2}{E_j}$$

n > 5 observations in each of J classifications distributed approximately as χ^2 with (J - 1) degrees of freedom

(12-4) Expected frequency of cell jk of a contingency table

$$E_{jk} = \frac{C_k}{n}(R_j)$$

R_j: marginal total of row j

C_k: marginal total of column k

(12-5) Test statistic for independence of the attributes of a population

$$X^2 = \sum_{j=1}^{J} \sum_{k=1}^{K} \frac{(O_{jk} - E_{jk})^2}{E_{jk}}$$

n > 5 observations in each of the J * K cells distributed approximately as χ^2 with (J - 1)(K - 1) degrees of freedom

ORGANIZED LEARNING QUIZ

1. When testing hypotheses involving one variance, the assumptions of the chi-square test include

    a. a random sample is taken from a normal population.
    b. the level of significance is .05.
    c. the sample size is over 30.
    d. no statements about the nature of the population.

2. When testing hypotheses involving one variance using the chi-square test,

    a. only a one-tail test can be used.
    b. only a two-tail test can be used.
    c. both a one-tail test and a two-tail test may be used, depending on the nature of the null hypothesis.
    d. neither a one-tail test nor a two-tail test is appropriate because of the nature of the null hypothesis.

3. The χ^2 distribution

    a. is similar to the binomial distribution.
    b. is symmetrical about the origin.
    c. is U shaped
    d. is nonnegative.

4. The χ^2 distribution

   a. has one parameter, its mean.
   b. has one parameter, its variance.
   c. has one parameter, its degrees of freedom.
   d. has one parameter, its standard deviation.

5. The chi-square test cannot be used for decisions

   a. as to whether two population variances are equal.
   b. goodness-of-fit tests.
   c. as to whether a population variance is equal to some number.
   d. as to whether a population variance is greater than some number.

6. In the goodness-of-fit test, the null hypothesis

   a. states that the variances of two distributions are equal.
   b. specifies some distribution.
   c. states that two means are equal.
   d. states that two means fit together but are not necessarily equal.

7. In the goodness-of-fit test, classifications should contain

   a. more than 30 observations.
   b. 5 or more observations.
   c. less observations than attributes.
   d. no more than 15 observations.

8. In testing for independence,

   a. even though a relationship exists in a sample, it might not exist in a population.
   b. if a relationship exists in a sample, it is sure to exist in the population.
   c. sampling cannot be done--the test must be done on the population.
   d. a different χ^2 distribution and set of tables are necessary, different from those used in goodness-of-fit tests.

9. The contingency table

   a. is helpful when testing hypotheses about a population variance.
   b. is the same as that used for the z test and the t test.
   c. can have as many rows and columns as needed.
   d. is restricted to a 2 x 2 size.

10. In a 2 x 2 contingency table with the marginal totals and the grand total given,

    a. any three cells must be given and then the fourth is fixed; thus there are three degrees of freedom.
    b. all four cells must be given; thus there are four degrees of freedom.
    c. two cells in one row or column must be given; thus there are two degrees of freedom.

d. knowing any one cell automatically determines the rest; thus there is only one degree of freedom.

11. In the test of statistical independence, the null hypothesis is rejected if the test statistic is very (large, small, large or small).

12. In the test of statistical independence and in the goodness-of-fit test, the analysis is based on the _____ of observations belonging to a particular classification, cell, or category.

13. If the null hypothesis is true, then the expected frequencies of each classification can be determined if the _____ is given.

14. Large differences between the observed frequencies and the expected frequencies will make the X statistic (large, small) and tend to (support, refute) the null hypotheses.

15. In testing hypotheses about a variance, if the null hypothesis states that the variance is equal to a constant, then a (one-tail, two-tail) test is used.

16. Explain why the X^2 test statistic is never negative.

_____

_____

_____

17. For a given level of significance, as the number of degrees of freedom increases, the critical χ^2 values become much larger than values of z that we are used to. Explain a probable reason why.

_____

_____

_____

18. Explain why the degrees of freedom of a contingency table are $(J - 1)(K - 1)$.

_____

_____

_____

19. Explain how a table of random digits could be tested as being random, using chi-square techniques.

_____

_____

20. What are the assumptions of the chi-square tests and what precautions must be taken when using the tests?

_____

_____

_____

ANSWERS TO ORGANIZED LEARNING QUIZ

1. a   2. c
3. d   4. c
5. a   6. b
7. b   8. a
9. c   10. d

11. large

12. frequency or number

13. sample size (grand total)

14. large, refute

15. two-tail

16. The test statistic is computed as the ratio of two items that are computed by summing squares; thus it can never be negative.

17. Since the chi-square distribution in nonnegative and dependent on sample size, its mode will move to the right as the sample size increases. Thus the critical values are placed far out on the right end.

18. The row and column totals and the grand total are known to determine the expected frequencies. Given these totals, all but one frequency in each row and in each column automatically determine the last one in each row and column; so one degree of freedom is lost in each row and each column.

19. If a group of digits is random, then one attribute is that each digit should occur 10% of the time. A goodness-of-fit test can be used to decide whether such is the case for a random sample from the table. Failing to reject the null hypothesis of no difference for this attribute does not completely show that a population is random, however. Other chi-square tests should be used as well on other known frequencies of occurrences on sets of known random numbers.

20. The test statistic for the one variance test follows the chi-square distribution with n - 1 degrees of freedom if the population is normal. For the goodness-of-fit and the independence tests, the test statistic only approximately follows a chi-square distribution. This situation occurs because the chi-square distribution is continuous and the

possible number of combinations of frequencies is discrete. Ensuring that there are at least five frequencies in each cell allows one to overlook this deficiency.

SAMPLE EXERCISE

The State Patrol has kept records of accidents that have occurred on a particular class of highways over one-week periods. They intend to use these data to show that a Poisson distribution can be used in simulation experiments concerning that particular class of highway. Use the goodness-of-fit procedure to test whether the records below are significantly different from a Poisson distribution with $\mu = 2.5$. Use a significance level of .05.

Highway Accident Statistics

Number of Accidents	Number of One-Week Periods Observed	Number Expected if Poisson Distributed ($\mu = 2.5$)
0	11	15
1	40	37
2	48	46
3	41	38
4	19	24
5	18	12
6	2	5
7	1	2

Calculations

O	E	O - E	$(O - E)^2$	$\frac{(O - E)^2}{E}$
11	15	-4	16	1.07777
40	37	3	9	.24324
48	46	2	4	.08696
41	38	3	9	.23684
19	24	-5	25	1.04167
18	12	6	36	3.00000
2	5	-3	9	1.80000
1	2	-1	1	.50000
Totals	180			7.97538

$X^2 = 7.97538$ with seven degrees of freedom. (Poisson mean is specified. If the mean had to be estimated, then one more degree of freedom would be lost.)

$$X^2_{7, .95} = 14.067 > 7.97538$$

Therefore we fail to reject and cannot discern a difference between the observed pattern and the Poisson.

CASE PROBLEM

Willamette Valley National Bank

Organized for over 100 years, with its home office in the largest population center of the state, Willamette Valley National Bank has grown to be one of the top 40 banks in the nation. Its branches number over 150 at present; and because of the large paperwork transfer between the branches and the centralized processing center, a large fleet of delivery vehicles must be maintained.

After noting the rise in number of accidents being reported during each month, Scott Burgner of the Motor Pool made plans to instigate a safety campaign for the drivers. At present over 70 drivers were in the system, representing differing age groups. Although he knew that the bank's insurance company representative paid rather close attention to the age distribution of his drivers, he wondered whether a relationship existed between number of accidents and age group. The insurance company gave him guidelines on the number of drivers in each age group he was allowed; and if he became overloaded in the younger groups, the overall rates were modified.

All accidents reported to the insurance company during the past 5 years were scanned and the following data were obtained. Because of driver turnover, there were records for 200 drivers.

Number of Accidents per Individual	Age Group				
	1	2	3	4	5
0	11	32	8	7	17
1	5	26	5	5	9
2	7	11	5	6	6
3 or more	9	11	6	6	8

Age groups were classified as

(1) under 20 years of age
(2) 20 to 25
(3) 26 to 30
(4) 31 to 40
(5) Over 40

COMPUTER PROGRAM

The flowchart and computer program that follow are written in BASIC and give the user a choice of one of the three chi-square tests studied in this chapter. Limits to inputs are 50 observations for the one variance test, 50 pairs of values for the goodness-of-fit test, and a maximum contingency table size of 10 x 10.

Sample runs are given for the Sample Exercise and for the Willamette Valley National Bank case.

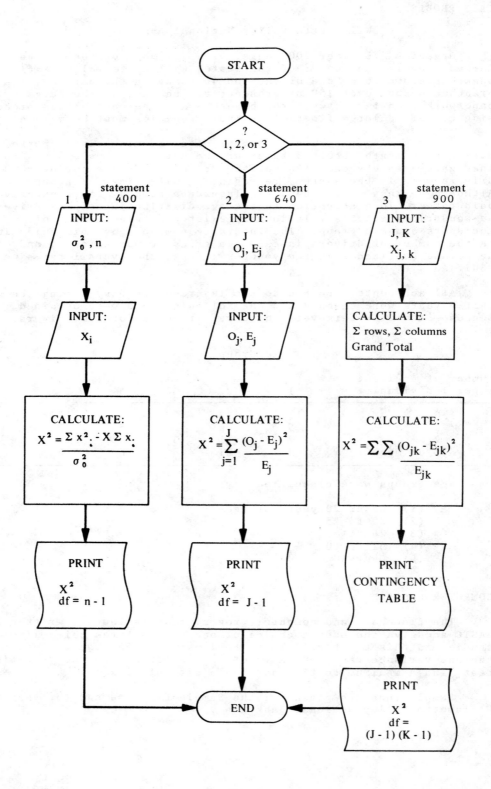

```
10 REM************************************************************
20 REM
30 REM       TITLE:     CHISQUAR
40 REM       LANGAUGE:  BASIC
50 REM       DESCRIPTIONS:
60 REM          THIS PROGRAM ALLOWS THE USER TO SELECT
70 REM          EITHER A ONE VARIANCE TEST; A GOODNESS OF
80 REM          FIT TEST; OR AN INDEPENDENCE TEST.  LIMITS
90 REM          TO INPUTS ARE 50 OBS FOR THE ONE VARIANCE
100 REM         TEST, 50 PAIRS OF VALUES FOR THE GOODNESS OF
110 REM         FIT TEST, AND A MAXIMUM CONTINGENCY TABLE
120 REM         SIZE OF 10 X 10.
130 REM
140 REM      INSTRUCTIONS:
150 REM         ENTER DATA AS REQUESTED BY THE PROGRAM.
160 REM
170 REM      PROGRAMMER:
180 REM         GROVER WM. RODICH, PORTLAND STATE UNIV.
190 REM
200 REM************************************************************
210 REM
220 KEY OFF
230 DIM X1(50),O3(10,10),E3(10,10),R(10),L(10),O(50),E(50)
240 CLS
250 PRINT "THIS PROGRAM GIVES YOU A CHOICE OF THREE CHI-SQUARE TESTS"
260 PRINT
270 PRINT "TYPE 0 AND RETURN TO EXIT PROGRAM"
280 PRINT "TYPE 1 AND RETURN FOR ONE VARIANCE TEST"
290 PRINT "TYPE 2 AND RETURN FOR GOODNESS OF FIT TEST"
300 PRINT "TYPE 3 AND RETURN FOR INDEPENDENCE TEST"
310 PRINT
320 PRINT "YOUR CHOICE";
330 INPUT A
340 CLS
350 IF A=0 THEN 1910
360 IF A=1 THEN 400
370 IF A=2 THEN 640
380 IF A<>3 THEN 280
390 GOTO 900
400 PRINT
410 PRINT "VARIANCE TEST"
420 PRINT "WHAT IS THE HYPOTHESIZED VARIANCE";
430 INPUT S
440 PRINT "WHAT IS THE SAMPLE SIZE";
450 INPUT N
460 PRINT "INPUT THE OBSERVATIONS ONE AT A TIME FOLLOWED BY A RETURN"
470 PRINT "FIRST ITEM";
480 INPUT X1(1)
490 PRINT "NEXT";
500 INPUT X1(2)
510 T2=X1(1)^2+X1(2)^2
520 T=X1(1)+X1(2)
530 FOR I=3 TO N
540 INPUT X1(I)
550 T=T+X1(I)
560 T2=T2+X1(I)^2
570 NEXT I
580 M=T/N
590 C=(T2-M*T)/S
600 PRINT
610 PRINT "THE CALCULATED CHI-SQUARE STATISTIC IS";C
620 PRINT "WITH";N-1;"DEGREES OF FREEDOM"
630 GOTO 1890
640 PRINT
650 PRINT "GOODNESS OF FIT TEST"
660 PRINT "HOW MANY CLASSES IN THE DATA?"
670 PRINT "J=";
680 INPUT J
690 PRINT
700 PRINT "INPUT EACH PAIR OF DATA ONE AT A TIME FOLLOWED BY A RETURN"
```

```
710 PRINT "TYPE THE OBSERVED FREQUENCY FIRST THEN THE EXPECTED"
720 PRINT "SEPARATE THE TWO WITH A COMMA"
730 PRINT "WHAT IS THE FIRST PAIR";
740 T=0
750 INPUT O(1),E(1)
760 T=T+((O(1)-E(1))^2)/E(1)
770 PRINT "NEXT PAIR";
780 INPUT O(2),E(2)
790 T=T+((O(2)-E(2))^2)/E(2)
800 IF J <3 THEN 860
810 PRINT "CONTINUE"
820 FOR I=3 TO J
830 INPUT O(I),E(I)
840 T=T+((O(I)-E(I))^2)/E(I)
850 NEXT I
860 PRINT
870 PRINT "THE CALCULATED CHI-SQUARE STATISTIC IS";T
880 PRINT "WITH";J-1;"DEGREES OF FREEDOM"
890 GOTO 1890
900 PRINT
910 PRINT "TEST OF STATISTICAL INDEPENDENCE"
920 PRINT "TYPE IN THE NUMBER OF CLASSES OF EACH ATTRIBUTE SEPARATED"
930 PRINT "BY A COMMA"
940 PRINT "THEN HIT RETURN"
950 PRINT "J AND K ARE";
960 INPUT J,K
970 PRINT "I WILL ASSUME THE CONTINGENCY TABLE HAS J=";J;"ROWS"
980 PRINT "AND K=";K;"COLUMNS"
990 PRINT
1000 PRINT "TYPE IN THE FIRST ROW OF DATA ONE AT A TIME"
1010 PRINT "FOLLOWED BY A RETURN"
1020 FOR Q=1 TO 10
1030 R(Q)=0
1040 L(Q)=0
1050 NEXT Q
1060 FOR I=1 TO K
1070 INPUT O3(1,I)
1080 NEXT I
1090 PRINT "CONTINUE ROW BY ROW FOR THE REST OF THE ITEMS."
1100 FOR Q=2 TO J
1110 FOR I=1 TO K
1120 INPUT O3(Q,I)
1130 NEXT I
1140 NEXT Q
1150 FOR Q=1 TO J
1160 FOR I=1 TO K
1170 R(Q)=R(Q)+O3(Q,I)
1180 NEXT I
1190 NEXT Q
1200 FOR I=1 TO K
1210 FOR Q=1 TO J
1220 L(I)=L(I)+O3(Q,I)
1230 NEXT Q
1240 NEXT I
1250 G=0
1260 FOR Q=1 TO J
1270 G=G+R(Q)
1280 NEXT Q
1320 FOR I=1 TO K
1330 FOR Q=1 TO J
1340 E3(Q,I)=R(Q)*L(I)/G
1350 NEXT Q
1360 NEXT I
1370 C=0
1380 FOR I=1 TO K
1390 FOR Q=1 TO J
```

```
1400 C=C+((O3(Q,I)-E3(Q,I))^2)/E3(Q,I)
1410 NEXT Q
1420 NEXT I
1430 PRINT
1440 PRINT
1450 PRINT TAB(25);"CONTINGENCY TABLE--OBSERVED"
1460 PRINT "CLASS";
1470 FOR I=1 TO K
1480 PRINT TAB(I*7);I;
1490 NEXT I
1500 PRINT TAB(I*7);"TOTAL"
1510 FOR Q=1 TO J
1520 PRINT Q;
1530 FOR I=1 TO K
1540 PRINT TAB(I*7);O3(Q,I);
1550 NEXT I
1560 PRINT TAB(I*7);R(Q)
1570 NEXT Q
1580 PRINT "TOTAL";
1590 FOR I=1 TO K
1600 PRINT TAB(I*7);L(I);
1610 NEXT I
1620 PRINT TAB(I*7);G
1630 PRINT
1640 PRINT
1650 PRINT TAB(25);"CONTINGENCY TABLE--EXPECTED"
1660 PRINT
1670 PRINT "CLASS";
1680 FOR I=1 TO K
1690 PRINT TAB(I*7);I;
1700 NEXT I
1710 PRINT TAB(I*7);"TOTAL"
1720 FOR Q=1 TO J
1730 PRINT Q;
1740 FOR I=1 TO K
1750 PRINT TAB(I*7);E3(Q,I);
1760 NEXT I
1770 PRINT TAB(I*7);R(Q)
1780 NEXT Q
1790 PRINT "TOTAL";
1800 FOR I=1 TO K
1810 PRINT TAB(I*7);L(I);
1820 NEXT I
1830 PRINT TAB(I*7);G
1840 PRINT
1850 PRINT
1860 PRINT "THE CALCULATED CHI-SQUARE STATISTIC IS";C
1870 PRINT
1880 PRINT "WITH";(J-1)*(K-1);"DEGREES OF FREEDOM"
1890 LOCATE 24,1:PRINT "TO CONTINUE PRESS ANY KEY";:A$=INPUT$(1)
1900 GOTO 240
1910 KEY ON:END
```

```
THIS PROGRAM GIVES YOU A CHOICE OF THREE CHI-SQUARE TESTS

TYPE 0 AND RETURN TO EXIT PROGRAM
TYPE 1 AND RETURN FOR ONE VARIANCE TEST
TYPE 2 AND RETURN FOR GOODNESS OF FIT TEST
TYPE 3 AND RETURN FOR INDEPENDENCE TEST

YOUR CHOICE? 2


GOODNESS OF FIT TEST
HOW MANY CLASSES IN THE DATA?
J=? 8

INPUT EACH PAIR OF DATA ONE AT A TIME FOLLOWED BY A RETURN
TYPE THE OBSERVED FREQUENCY FIRST THEN THE EXPECTED
SEPARATE THE TWO WITH A COMMA
WHAT IS THE FIRST PAIR? 11,15
NEXT PAIR? 40,37
CONTINUE
? 48,46
? 41,38
? 19,24
? 18,12
? 2,5
? 1,2

THE CALCULATED CHI-SQUARE STATISTIC IS 7.975375
WITH 7 DEGREES OF FREEDOM


TO CONTINUE PRESS ANY KEY


THIS PROGRAM GIVES YOU A CHOICE OF THREE CHI-SQUARE TESTS

TYPE 0 AND RETURN TO EXIT PROGRAM
TYPE 1 AND RETURN FOR ONE VARIANCE TEST
TYPE 2 AND RETURN FOR GOODNESS OF FIT TEST
TYPE 3 AND RETURN FOR INDEPENDENCE TEST

YOUR CHOICE? 3

TEST OF STATISTICAL INDEPENDENCE
TYPE IN THE NUMBER OF CLASSES OF EACH ATTRIBUTE SEPARATED
BY A COMMA
THEN HIT RETURN
J AND K ARE? 4,5
I WILL ASSUME THE CONTINGENCY TABLE HAS J= 4 ROWS
AND K= 5 COLUMNS

TYPE IN THE FIRST ROW OF DATA ONE AT A TIME
FOLLOWED BY A RETURN
? 11
? 32
? 8
? 7
? 17
```

```
CONTINUE ROW BY ROW FOR THE REST OF THE ITEMS.
? 5
? 26
? 5
? 5
? 9
? 7
? 11
? 5
? 6
? 6
? 9
? 11
? 6
? 6
? 8
```

CONTINGENCY TABLE--OBSERVED

CLASS	1	2	3	4	5	TOTAL
1	11	32	8	7	17	75
2	5	26	5	5	9	50
3	7	11	5	6	6	35
4	9	11	6	6	8	40
TOTAL	32	80	24	24	40	200

CONTINGENCY TABLE--EXPECTED

CLASS	1	2	3	4	5	TOTAL
1	12	30	9	9	15	75
2	8	20	6	6	10	50
3	5.6	14	4.2	4.2	7	35
4	6.4	16	4.8	4.8	8	40
TOTAL	32	80	24	24	40	200

THE CALCULATED CHI-SQUARE STATISTIC IS 9.675497

WITH 12 DEGREES OF FREEDOM
TO CONTINUE PRESS ANY KEY

```
THIS PROGRAM GIVES YOU A CHOICE OF THREE CHI-SQUARE TESTS

TYPE 0 AND RETURN TO EXIT PROGRAM
TYPE 1 AND RETURN FOR ONE VARIANCE TEST
TYPE 2 AND RETURN FOR GOODNESS OF FIT TEST
TYPE 3 AND RETURN FOR INDEPENDENCE TEST

YOUR CHOICE? 0
OK
```

13

ANALYSIS OF VARIANCE

CHAPTER LEARNING OBJECTIVES

On completing this chapter, the student should be able to

- Explain the use of the F statistic and describe the conditions where it can be used as an extension of the T statistic.

- Discuss the test of the hypothesis stating that the two unknown variances of two normal populations are equal.

- Describe the process of partitioning sums of squares into components in order to estimate variation due to different sources.

- Explain why a collection of statistical methods used to test hypotheses about the equality of *means* is called the analysis of *variance*.

- Design an experiment and organize the data to investigate several treatment effects.

- Design an experiment and organize the data to investigate the several effects of the various levels of two treatments.

- Discuss alternatives available to the experimenter when the null hypothesis of equal means is rejected.

- List the assumptions present when using analysis of variance.

SYNOPSIS

In Chapter 11 a method was developed to test the equality of two population means by investigating data sampled randomly from the two populations. Analysis of variance is a collection of statistical methods used to study observations from two *or more* populations. The null hypothesis and its alternative in Chapter 11 were that the two population means are equal or they are not equal. In this chapter we are interested in whether all the population means are equal or whether there are at least two means that are not equal. Unfortunately, analysis of variance does not distinguish the populations with significantly different means.

Examples of problems in which we would want to use analysis of variance are easily found by an extension of the problems in Chapter 11. One situation would be the test of durability of three or more different brands of lightbulbs. Another is a test of five brands of tires instead of two to see if their average life is the same.

The f Distribution

An f distribution is used whenever a test about two variances is conducted. It is the distribution of the ratio of two unbiased sample variances, assuming the samples were taken from normally distributed populations. Since the ratio of two variances is non-negative, the f distribution is never negative. The shape of the f distribution depends on the values of its two parameters, ν_1 and ν_2, where ν_1 is ($n_1 - 1$) from the numerator of the ratio and ν_2 is ($n_2 - 1$) from the denominator. The test statistic, F, is computed by measuring the quotient of two actual sample variances, s_1^2 and s_1^2.

Since the f distribution has two parameters, f tables are two dimensional with a whole table needed for each level of significance. By always placing the larger variance in the numerator, F will always be equal to or greater than one. The table values are then the critical f values, which cuts off the upper tail equal to α of the distribution. If the null hypothesis is that two population variances are equal, then rejection will occur if the differences in their sample variances cause their ratio to be in the cutoff region.

Testing for The Difference Between Two Variances

The assumptions for the test are that random samples are drawn from the two populations in question and that the populations are normal. The population variances are not known, but the unbiased sample variances can be computed. The F statistic is the appropriate statistic to test the following three sets of null hypotheses and their alternates.

$$H_0: \sigma_1^2 = \sigma_2^2 \qquad H_0: \sigma_1^2 \geq \sigma_2^2 \qquad H_0: \sigma_1^2 \leq \sigma_2^2$$
$$H_1: \sigma_1^2 \neq \sigma_2^2 \qquad H_1: \sigma_1^2 < \sigma_2^2 \qquad H_1: \sigma_1^2 > \sigma_2^2$$

The first case is a two-tailed test and both small and large values of F may cause rejection. However, by always putting the larger sample variance in the numerator of the F ratio, the two-tailed test can be done by using the one-tailed test method. The critical or rejection region is determined by dividing the level of significance by two ($\alpha/2$).

One-Way Analysis of Variance

The f distribution allows an investigator to extend the tests concerning population means from the two population tests studied with the t distribution in Chapter 11 to three or more population tests.

Analysis of Variance--The Rationale. It may seem strange to students on their first encounter with analysis of variance that variances are investigated when the hypothesis concerns the equality of means of the populations. It is not the variances of the populations that are of interest (they are assumed equal) but the appropriate variances created by breaking down the total variation of all data into separate components. The partitioned variations (sums of squares), when averaged by appropriate constants (degrees of freedom), should have compatible values, since they are all unbiased estimates of the total population variance.

A ratio of these variances against a pure error variance follows an F distribution if the null hypothesis of equal means is true. If the means are not equal, then the average value of the variation between levels is increased, which tends to produce a larger F value. Values of F larger than tabular values for the given degrees of freedom for the numerator and denominator lead to the rejection of the null hypothesis and the conclusion is drawn that not all the means of the level populations are equal.

In a one-way classification each population sample being tested is said to be the result of a level of treatment; therefore if k populations are being considered, there are k levels. Each sample is made up of units that have been randomly assigned to any of the k samples; for this reason, the design is sometimes called a completely randomized design.

Two-Way Analysis of Variance

In a two-way classification two treatments are investigated, each with two or more levels. Treatments are also called factors in some textbooks and thus the experiment is called a two-level factorial design. An example might be the investigation of both the training of workers and types of machines used in the production and productivity of a metal product component. Levels of training, different types of machines, and their interactive effects can be compared.

When using analysis-of-variance results, the assumptions that must be made are that the k populations are normally distributed and that they all have the same population variance. Slight modifications of these assumptions are not too noticeable in the decision to accept or reject and even less noticeable when the sample sizes are kept equal. This is a further incentive for the designer of the experiment to try to keep equal sample sizes.

KEY TERMS

Analysis of variance — A body of statistics used to investigate whether the data from two or more samples are enough alike that it could be conclud-

	ed the samples could all have been drawn from a single population.
ANOVA table	A table that conventionally lists the parameters used in the development of the experimental F statistic.
Completely randomized design	If when designing an experiment the units are assigned at random to any one of the levels of treatments under investigation, the classification is referred to as a completely randomized design. Other designs include randomized blocks and Latin squares.
Fixed-effects model	Any conclusions reached at the end of the experiment are valid only for the populations of the levels of treatments used for investigation. If the levels were selected at random from a larger set of levels, then the experiment is called a random-effects model.
Partitioning sums of squares	The algebraic separation of the total sum of squares (total variation) into components.
Sum of squares	Another term for variation. The deviation of each observation from the mean is computed, squared, and then all of them are summed.
Sum of squares of error	The pooled variation within the levels of treatments due to chance or sampling error.
Sum of squares of treatments	The variation between the levels of treatments due to chance and also possibly due to the differences of the levels. In a two-way analysis of variance this variation is divided into three distinct parts: the variation due to the first factor; the variation due to the second factor; and the variation due to the interaction of the two factors that is not attributable to either of the two factors alone.
Total sum of squares	The variation of all observations without regard to their level of treatment. The data are combined and regarded as one large sample.
Treatment	A qualitative independent variable causing particular reponses of a dependent variable. Levels of treatments refer to the different settings of the variable. These might involve different brands, a different time or place, or different environmental surroundings.

FORMULA REVIEW

The formulas and calculations used in completing an analysis of variance are best displayed in tabular fashion. Statisticians have established the conventional ANOVA table to recap the relevant variables necessary to compute the value of the F statistic. Figure 1 gives the general form of a one-way analysis, whereas Figure 2 gives the proper form for a two-way ANOVA. Figure 3 shows a graphical description of how the total sum of squares is partitioned when two-way analysis of variance is used.

Source of Variation	SS	df	MS	F
Between levels	SSB	K - 1	MSB = SSB/(K - 1)	$\frac{MSB}{MSW}$
Within levels	SSW	N - K	MSW = SSW/(N - K)	
Total	SST	N - 1		

Figure 1   One-Way ANOVA Table

Source of Variation	SS	df	MS	F
Treatment A (row)	SSR	J - 1	MSR = SSR/(J - 1)	MSR/MSE
Treatment B (column)	SSC	K - 1	MSC = SSC/(K - 1)	MSC/MSE
Interaction	SSI	(J-1)(K-1)	MSI = SSI/(J-1)(K-1)	MSI/MSE
Error	SSE	JK(n-1)	MSE = SSE/JK(n-1)	
Total	SST	JKn-1		

Figure 2   Two-Way ANOVA Table

Total sum of squares	SST			
One-Way ANOVA:	SSB			SSW
Two-Way ANOVA:	SST_r			
	SSR	SSC	SSI	SSE

Figure 3   Partitioning Total Sum of Squares

ORGANIZED LEARNING QUIZ

1. The f distribution

   a. is a normal distribution.
   b. is a one-parameter distribution.
   c. is a two-parameter distribution.
   d. is an approximation to the binomial distribution.

2. When independent random samples are drawn from two normal populations with equal variances, then s_1^2/s_2^2 follows

a. a constant value.
   b. no known pattern.
   c. a normal distribution.
   d. an f distribution.

3. The shape of the f distribution

   a. is symmetrical.
   b. is nonsymmetrical.
   c. has a constant negative slope.
   d. has a constant positive slope.

4. The shape of the f distribution depends upon

   a. the degrees of freedom of s_1^2 and s_2^2.
   b. whether the mean is greater than one.
   c. the table one is using.
   d. the value of α.

5. Critical values for the F statistic

   a. will always be in the upper tail.
   b. can be in the upper tail or the lower tail.
   c. will always be in the lower tail.
   d. are within one standard deviation of the mean.

6. If the larger sample variance is always designated as the numerator of the F statistic, then

   a. the value of α used in a F table is always .10.
   b. the F statistic converts to chi-square.
   c. the critical region will always be in the lower tail.
   d. the test will always be a one-tailed test.

7. When testing for the comparison of two population means,

   a. the binomial test is a good approximation.
   b. both a t test and an f test will give identical results.
   c. no test is needed. If the population means are equal, then the sample means will be equal.
   d. a ratio of sample means will follow the f distribution.

8. A fixed-effects model allows us to make inferences about

   a. the k populations being investigated.
   b. all the populations from which the k populations are drawn.
   c. the validity of the questionnaire used.
   d. the total sums of squares.

9. A treatment in analysis of variance

   a. refers to a level of a qualitative independent variable.
   b. refers to the particular statistical test used.
   c. refers to a single independent variable with two or more levels.
   d. is not an analysis-of-variance term.

10. Two-way analysis of variance

      a. involves two levels of a treatment.
      b. involves two or more levels of two treatments.
      c. involves the testing of only one null hypothesis.
      d. involves the testing of only two null hypotheses.

11. t^2 is equal to f when _____.

12. MSB and MSW are unbiased estimates of _____.

13. SSW is the result of _____ the sums of squares of deviations from the individual samples.

14. As treatment level means get farther apart, the larger the value _____.

15. _____ values of F lead to a rejection of the null hypothesis.

16. _____ is the name of an analysis-of-variance table where the calculations are displayed.

17. The sums of squares within the levels of a treatment are called _____.

18. Total sum of squares are partitioned into _____ and _____.

19. If a five-level one-way analysis of variance leads to the rejection of the null hypothesis of equal means, level two could be compared with level four, using a _____ test.

20. The degrees of freedom of SST are always _____.

21. Explain the difference between variance and variation.

    _____
    _____
    _____

22. Explain what variances are investigated in analysis of variance.

    _____
    _____
    _____

23. Explain why the f test is necessary when comparing more than two means.

    _____
    _____
    _____

24. Explain further test possibilities when the f test leads to a rejection.

   _____

   _____

   _____

25. Explain the assumptions of analysis of variance.

   _____

   _____

   _____

ANSWERS TO ORGANIZED LEARNING QUIZ

1. c
2. d
3. b
4. a
5. b
6. d
7. b
8. a
9. c
10. b

11. The degrees of freedom of the numerator equal one.

12. the population variance

13. pooling

14. SSB

15. large

16. ANOVA

17. SSW

18. SSB and SSW

19. t test or a two-level analysis of variance

20. N - 1

21. Variation is the sum of the squares of the deviations of each observation from its sample mean. Variance is the average variation obtained by dividing the variation by its appropriate degrees of freedom.

22. The variance between the sample means is compared to the pooled variance within the samples.

23. The t test can only be used when comparing two sample means.

24. The Tukey method, Scheffe's method, t test (pairwise comparisons), or visual inspection.

25. Random samples drawn from normal populations with equal population variances.

SAMPLE EXERCISE

A business statistics professor teaches first-year statistics to three large classes each term. Last term he randomly chose days to check attendance in each of his three classes. The number of absences were noted and recorded in the table. Test the hypothesis that the mean absences of each class are the same.

The large number of calculations in both one-way and two-way analysis of variance are best made and kept track of by use of the calculation table introduced in earlier chapters. By a slight addition of total means and variation in the one-way table and row and column totals in the two-way table, all necessary calculations are available for use in the final ANOVA table. The computation table and its explanation for the Sample Exercise follows.

Computation Table   One-Way Analysis of Variance

Observation		Levels of Treatment			X_1^2	X_2^2	X_3^2		
		X_1	X_2	X_3					
	1	5	2	4	25	4	16		
	2	2	3	5	4	9	25		
	3	4	3	4	16	9	16		
	4	4	4	6	16	16	36		
	5	3	4	3	9	16	9		
	6	6	2	6	36	4	36		
	7	4		7	16		49		
	8	4		5	16		25		
	9			5			25		
								Grand Totals	
Totals		Ⓐ 32	Ⓐ 18	Ⓐ 45	Ⓑ 138	Ⓑ 58	Ⓑ 237	Ⓕ 95	Ⓖ 433
Means and Corrections		Ⓒ 4	Ⓒ 3	Ⓒ 5	Ⓓ 128	Ⓓ 54	Ⓓ 225	Ⓗ 4.13	Ⓘ 392.4
Variations					Ⓔ 10 +	Ⓔ 4 +	Ⓔ 12		Ⓙ SS = 40.6
						Ⓚ SSE = 26			

159

Explanation of Table

- (A) The sum of observations at each level: $\sum_{i=1}^{n_i} X_{ik}$ for each k

- (B) The sum of observations squared: $\sum_{i=1}^{n_i} X_{ik}^2$ for each k

- (C) The mean of observations for each level: (A)/n_k

- (D) The correction factor to reduce the sum of squares of deviations from the origin to the sum of squares of deviations from the sample mean: (A) * (C)

- (E) The variation of each level (sum of squares of deviation from sample mean): (B) - (D)

- (F) The total of all observations from all levels: $\sum^{k}$ (A)

- (G) The total of all observations after squaring: $\sum^{k}$ (B)

- (H) The mean of all observations: (F)/N

- (I) The correction factor to reduce the sum of squares of deviations from the origin to the sum of squares of deviations from the overall mean: (F) * (H)

- (J) The total variation: (G) - (I)

- (K) The pooled variation: $\sum^{k}$ (E)

Source of Variation	SS	df	MS	F
Between levels	14.6	2	7.3	$\frac{7.3}{1.3} = 5.62$
Within levels	26	20	1.3	
Total	40.6	22		

Figure 4 ANOVA Table for the Business Statistics Professor

For an alpha value of .05, the table value for F with 2 degrees of freedom in the numerator and 20 degrees of freedom in the denominator is 3.49. Thus the professor would conclude that there is no difference in the absenteeism among his classes, since the calculated F (5.62) is larger than the table value (3.49).

CASE PROBLEM

Cropco, Inc.

To aid in merchandising and cost effectiveness, Cropco, Inc., a grain exporter, decided to acquire its own computer system. The choice is limited to four systems, all compatible with present and future requirements and having negligible cost differences. Services of one of three people will be contracted to perform programming. All have equal qualifications, will perform identical operations within a specified time, and receive equal pay. Mr. Olds, manager of Cropco, wonders if there might be differences in programming abilities and machine performance ratings under an actual production application. The decision on which system to acquire and which person to contract would be affected if any system or any programmer stands out as statistically different from the rest.

Many similar problems involving analysis and forecasting are obtained from a major university. Four of these problems were randomly assigned to each programmer/computer combination to insure a completely randomized two-factor design. Arrangements have been made with the four manufacturers to allow each of the programmers to run their programs on the respective computer.

Program entry time is monitored and accuracy of the run noted. Programming time, personal machine operation, and machine performance were rated on a scale of 1 to 10. The data are as follows.

Person	Machine 1	Machine 2	Machine 3	Machine 4
1	4 3 1 2	4 2 1 1	3 2 3 2	2 1 3 4
2	3 1 1 2	4 3 3 4	4 6 6 7	4 3 3 4
3	3 4 3 1	6 6 5 7	3 6 5 6	6 5 6 3

The assumptions of analysis of variance that Mr. Olds must make are (1) the 12 cells represent 12 random samples taken from 12 populations, (2) the 12 populations are all normal, and (3) the variances of the 12 populations are equal. The three main null hypothesis of interest (and their alternates) are

1. H_0: there is no difference between the programmers
   H_1: the programmers' performances are not all the same

2. H_0: there is no difference between machines
   H_1: the machines' performances are not all the same

3. H_0: there are no interaction effects between the programmers and the machines

H_1: there is an interaction effect between factors

Source of Variation	SS	df	MS	F	Table F
Treatment A	42.875	2	21.44	17.02	3.28
Treatment B	27.896	3	9.30	7.38	2.88
Interaction AB	27.792	6	4.63	3.67	2.38
Error	45.25	36	1.26		
Total	143.813	47			

Figure 5   ANOVA Table for Cropco, Inc.

The Table F values given in Figure 5 are for an alpha of .05. Mr. Olds cannot conclude there is a statistically significant difference in the performance of the machines or of the programmers.

The computations necessary for completion of the ANOVA table are grouped in a computation table below.

Computation Table Two-Way Analysis of Variance

A \ B	1		2		3		4		Totals	
	X	X^2	X	X^2	X	X^2	X	X^2		
1	4	16	4	16	3	9	2	4	Ⓟ 1444	
	3	9	2	4	2	4	1	1		
	1	1	1	1	3	9	3	9	Ⓔ 38	Ⓗ 108
	2	4	1	1	2	4	4	16		
	10 Ⓚ 30		8 Ⓚ 22		10 Ⓚ 26		10 Ⓚ 30		Ⓛ 364	
	100		64		100		100			
	X	X^2	X	X^2	X	X^2	X	X^2		
2	3	9	4	16	4	16	4	16	Ⓟ 3364	
	1	1	3	9	6	36	3	9		
	1	1	3	9	6	36	3	9	Ⓔ 58	Ⓗ 252
	2	4	4	16	7	49	4	16		
	7 Ⓚ 15		14 Ⓚ 50		23 Ⓚ 137		14 Ⓚ 50		Ⓛ 970	
	49		196		529		196			
	X	X^2	X	X^2	X	X^2	X	X^2		
3	3	9	6	36	3	9	6	36	Ⓟ 5625	
	4	16	6	36	6	36	5	25		
	3	9	5	25	5	25	6	36	Ⓔ 75	Ⓗ 393
	1	1	7	35	6	36	3	9		
	11 Ⓚ 35		24 Ⓚ 146		20 Ⓚ 106		20 Ⓚ 106		Ⓛ 1497	
	121		576		400		400			
Totals	Ⓔ 28	Ⓗ 80	Ⓔ 46	Ⓗ 218	Ⓔ 53	Ⓗ 269	Ⓔ 44	Ⓗ 186	Ⓕ 171	Ⓘ 753
	Ⓡ 784		Ⓡ 2116		Ⓡ 2809		Ⓡ 1936		Ⓜ 2831	

Overall mean = 3.5625 Ⓖ

Ⓐ $n = 4$ Ⓙ $143.8125 = SST$
Ⓑ $a = 3$ Ⓝ $98.5625 = SST_r$
Ⓒ $b = 4$ Ⓞ $45.25 = SSE$
Ⓓ $N = 48$ Ⓠ $42.875 = SSR$
                Ⓢ $27.8958\overline{3} = SSC$
                Ⓣ $27.7916\overline{6} = SSI$

Explanation of Table

- (A) The sample size of each cell
- (B) The number of levels of treatment (factor) A
- (C) The number of levels of treatment (factor) B
- (D) The overall number of observations
- (E) The total of all cells in each row (or column)
- (F) The overall total of all cells: Σ(E)
- (G) The overall mean: (F)/N
- (H) The total sum of squares of observations in each row (column)
- (I) The overall total of the squared observations
- (J) The corrected total sum of squares of deviations of all observations from the overall mean: (I) - (F) * (G)
- (K) The square of the cell totals
- (L) The sum of the squares of the cell totals for each row: Σ(K)
- (M) The overall sum of squares of the cell totals: Σ(L)
- (N) The corrected sum of squares for treatments: (M/n) - (F) * (G)
- (O) The sums of squares for error: (J) - (N)
- (P) The square of the sum of the cell totals for each row: (E) * (E)
- (Q) The sum of squares measuring the variability of the sample means obtained for various levels of A: (Σ(P)/bn) - (F) * (G)
- (R) The square of the sum of the cell totals for each column: (E) * (E)
- (S) The sum of squares measuring the variability of the sample means obtained for various levels of B: (Σ(R)/an) - (F) * (G)
- (T) The sum of squares of interaction measuring the variability obtained by A and B together that are not attributable to either A or B alone: (N) - (Q) - (S)

COMPUTER PROGRAM

The BASIC program given will print the computation table for one-way analysis of variance. The dimension statement in line 140 limits the program to 20 levels with a maximum sample size of 20. An output is given for the business statistics professor problem.

```
10 REM****************************************************************
20 REM
30 REM     TITLE:     ONEWAY
40 REM     LANGAUGE:  BASIC
50 REM     DESCRIPTIONS:
60 REM        THE PROGRAM WILL PRINT THE COMPUTATION TABLE
70 REM        FOR ONE-WAY ANALYSIS OF VARIANCE.
80 REM     INSTRUCTIONS:
90 REM        ENTER DATA AS REQUESTED BY THE PROGRAM.
100 REM    PROGRAMMER:
110 REM       GROVER WM. RODICH, PORTLAND STATE UNIV.
120 REM
130 REM****************************************************************
140 DIM X(20,20),N(20),T(22),M(22),V(22)
150 KEY OFF
160 CLS
170 PRINT "PROGRAM FOR ONEWAY ANALYSIS OF VARIANCE"
180 PRINT
190 PRINT "     INPUT 0 TO EXIT PROGRAM"
200 PRINT "     INPUT 1 TO CONTINUE PROGRAM"
210 PRINT
220 PRINT "        YOUR CHOICE";
230 INPUT G: IF G=0 THEN 1340
240 IF G<>1 THEN 190
250 FOR I=1 TO 22
260 V(I)=0
270 NEXT I
280 CLS
290 PRINT "HOW MANY LEVELS OF THE TREATMENT";
300 INPUT K
310 FOR I=1 TO K
320 PRINT "HOW MANY ITEMS IN LEVEL NO.";I;
330 INPUT N(I)
340 NEXT I
350 T1=0
360 T2=0
370 FOR I=1 TO K
380   T1=T1+N(I)
390 NEXT I
400 FOR I=1 TO 20
410 FOR J=1 TO 20
420 X(I,J)=0
430 NEXT J
440 NEXT I
450 FOR I=1 TO K
460 PRINT "PLEASE INPUT THE RESPONSES IN LEVEL ";I
470 FOR J=1 TO N(I)
480 INPUT X(J,I)
490 NEXT J
500 NEXT I
510 PRINT
520 PRINT
530 PRINT
540 PRINT
550 PRINT
560 FOR I=K+1 TO 2*K
570 FOR J=1 TO N(I-K)
580 X(J,I)=X(J,I-K)*X(J,I-K)
590 NEXT J
600 NEXT I
610 FOR I=1 TO K*2+2
620 T(I)=0
630 NEXT I
640 FOR I=1 TO K
650 N(K+I)=N(I)
660 NEXT I
670 FOR I=1 TO 2*K
680 FOR J=1 TO N(I)
690 T(I)=T(I)+X(J,I)
700 NEXT J
710 NEXT I
720 FOR I=1 TO K
730 M(I)=T(I)/N(I)
740 NEXT I
750 FOR I=K+1 TO 2*K
760 M(I)=(M(I-K)*T(I-K))
770 V(I)=T(I)-M(I)
780 NEXT I
```

```
790 T(2*K+I)=0
800 T(2*K+2)=0
810 T(2*K+1)=0
820 T(2*K+2)=0
830 FOR I=1 TO K
840 T(2*K+1)=T(2*K+1)+T(I)
850 T(2*K+2)=T(2*K+2)+T(K+I)
860 NEXT I
870 M(2*K+1)=T(2*K+1)+T(I)
880 M(2*K+1)=T(2*K+1)/T1
890 M(2*K+2)=M(2*K+1)*T(2*K+1)
900 V(2*K+2)=T(2*K+2)-M(2*K+2)
910 FOR I=1 TO 20
920 PRINT I;
930 E=0
940 FOR J=1 TO K*2
950 IF X(I,J)<>0 THEN 980
960 IF E=K*2-2 THEN 1020
970 E=E+1
980 PRINT TAB(J*8);X(I,J);
990 NEXT J
1000 PRINT
1010 NEXT I
1020 PRINT
1030 PRINT
1040 PRINT
1050 PRINT "TOT";
1060 FOR I=1 TO 2*K+2
1070 PRINT TAB(I*8);T(I);
1080 NEXT I
1090 PRINT
1100 PRINT "M&C";
1110 FOR I=1 TO 2*K+2
1120 J=I*10
1130 PRINT TAB(I*8);M(I);
1140 NEXT I
1150 PRINT
1160 E=0
1170 PRINT
1180 PRINT "SSE=";
1190 FOR I=K+1 TO 2*K-1
1200 E=E+V(I)
1210 PRINT V(I);"+";
1220 NEXT I
1230 E=E+V(2*K)
1240 PRINT V(2*K);"=";E
1250 PRINT
1260 PRINT "SST=";V(2*K+2)-E
1270 PRINT
1280 PRINT "SS=";
1290 PRINT V(2*K+2)
1300 PRINT
1310 PRINT
1320 LOCATE 24,1:PRINT "TO CONTINUE PRESS ANY KEY";:A$=INPUT$(1)
1330 GOTO 160
1340 KEY ON:END
```

```
PROGRAM FOR ONEWAY ANALYSIS OF VARIANCE

        INPUT 0 TO EXIT PROGRAM
        INPUT 1 TO CONTINUE PROGRAM

        YOUR CHOICE? 1


HOW MANY LEVELS OF THE TREATMENT? 3
HOW MANY ITEMS IN LEVEL NO. 1 ? 8
HOW MANY ITEMS IN LEVEL NO. 2 ? 6
HOW MANY ITEMS IN LEVEL NO. 3 ? 9
PLEASE INPUT THE RESPONSES IN LEVEL   1
? 5
? 2
? 4
? 4
? 3
? 6
? 4
? 4
PLEASE INPUT THE RESPONSES IN LEVEL   2
? 2
? 3
? 3
? 4
? 4
? 2

PLEASE INPUT THE RESPONSES IN LEVEL   3
? 4
? 5
? 4
? 6
? 3
? 6
? 7
? 5
? 5
```

1	5	2	4	25	4	16		
2	2	3	5	4	9	25		
3	4	3	4	16	9	16		
4	4	4	6	16	16	36		
5	3	4	3	9	16	9		
6	6	2	6	36	4	36		
7	4	0	7	16	0	49		
8	4	0	5	16	0	25		
9	0	0	5	0	0	25		
10	0	0	0	0				
TOT	32	18	45	138	58	237	95	433
M&C	4	3	5	128	54	225	4.130435	
								392.3913

SSE= 10 + 4 + 12 = 26

SST= 14.60867

SS= 40.60867

TO CONTINUE PRESS ANY KEY

PROGRAM FOR ONEWAY ANALYSIS OF VARIANCE

        INPUT 0 TO EXIT PROGRAM
        INPUT 1 TO CONTINUE PROGRAM

        YOUR CHOICE? 0
OK

14

ELEMENTS OF MODERN DECISION THEORY

CHAPTER LEARNING OBJECTIVES

On completing this chapter, a student should be able to

- Construct and use a decision matrix as an aid to decision making among alternatives.

- Classify decision making as being under certainty, under uncertainty, or under conflict.

- Determine the choice a decision maker would make using both a maximax rule and a minimax rule when making decisions under uncertainty.

- Calculate the expected gain or loss of a course of action when probabilities are attached to states of nature.

- Use Bayesian strategy to minimize expected opportunity losses while making decisions of alternative courses of action.

- Calculate the value of additional information.

- Determine the game strategy that players should use when using mixed strategies and be able to determine the value of the game to each player.

SYNOPSIS

In this chapter some new techniques of statistical decision making are examined. Hypothesis testing and parameter interval estimation, discussed in earlier chapters, are generally referred to as the classical approach to statistics. The Bayesian approach to statistics utilizing decisions based on posterior probabilities has also recently become a part of the total package of decision theory available to help decision makers be more effective. To begin the study of the Bayesian approach, a matrix of payoffs called the decision matrix is studied.

The Decision Matrix

In preceding chapters the decision maker structured his problem with just two alternatives--reject or don't reject the null hypothesis. In this section more than two alternatives are considered, along with the consequences or payoffs--either gains or losses--for each alternative. The assumptions are that the payoffs for each alternative are known and that the decision maker is a rational being whose preference of maximum gain (or minimum losses) guides his actions.

Nature is also a partner in each decision result; often situations develop that the decision maker cannot control. A two-dimensional table listing the results of the combinations of the courses of actions of the decision maker and the states of nature is labeled a decision matrix. If the future state of nature is known, then the matrix degenerates into a single row with the cells listing the payoff of each decision alternative with that one state of nature. This process is called decision making under certainty and poses no real problem to the decision maker. Decision making under uncertainty results when the future state of nature is not known and several possibilities exist for it. If the states of nature are under control of an intelligent adversary, then the decision maker faces decision making under conflict.

Decision Making under Uncertainty

Maximax Rule. An extreme optimist would look at all the best payoffs of each state of nature and assume that the state of nature with the highest payoff cell will occur. He is choosing the course of action that will result in the best of the best; if successful, he is said to be following the maximax rule. If there is a tie for the maximax payoff, then he is ambivalent as to which course of action he chooses as long as one of the maximax payoffs is a possibility for that action.

Minimax Rule. If the payoffs are expressed as losses incurred instead of gains, then the decision maker would apply opposite reasoning to that expressed in the preceding paragraph. By noting the maximum loss possible for each course of action, the action having the smallest value of the maximums is chosen. This is called the minimax rule.

Bayesian Rule. By giving the decision maker more information, we decrease the uncertainty surrounding his decision. Establishing probabilities for the states of nature will certainly make him more confident, although he is still lacking complete knowledge of the future state of nature.

Expected payoff. When probabilities are attached to the states of nature, it is possible to calculate the long-run expectation of gain or loss from each course of action. The choice is then the course of action with the largest gain (or smallest loss). This decision choice is referred to as the Bayesian action and the process as the Bayesian decision rule.

Opportunity loss. If the decision maker could "Monday morning quarterback" his decision by looking back from the future after the decision has been made and the state of nature has occurred, he would sometimes be satisfied and sometimes have regrets. For each state of nature that might have occurred, his reget would be the opportunity lost by not selecting the best course of action. These regrets are then calculated for the whole matrix by subtracting each item in a state of nature row from the largest entry in the row. The Bayesian decision rule can be applied to the revised matrix and the decision maker will select the course of action with the minimum expected opportunity loss.

Bayes' Theorem

The probabilities used in the Bayesian decision rule are subjective a priori probabilities. Even though based on excellent historical evidence, they are still made prior to the outcome of the decision. If additional information can be obtained, it may be possible to update the prior probabilities and further reduce the uncertain environment of the decision. The updated probabilities are called posterior probabilities, although still being derived before the state of nature occurs. Bayes' theorem (law or rule) is employed to calculate the posterior probabilities from the prior ones and the new information gained.

The tabular approach to generating the conditional probabilities utilizes a joint probability table in which the cells are the joint probabilities and the simple or marginal probabilities are in the margins. The posterior conditional probabilities are derived by dividing the joint probabilities by the appropriate marginal probability.

The derived conditional probabilities are in reverse of known conditional probabilities. For instance, if we know the probability of our neighbor carrying an umbrella, if it rains, we may be able to derive the probability of it raining if he carries an umbrella. Certainly the rainmaker does not check with our neighbor before turning on a downpour, but the probabilities derived are useful for decision-making purposes.

The Bayes Strategy

The foregoing discussion of Bayes decision rule and Bayesian actions centered on the revision of the prior probabilities assigned to the various possible states of nature. These probabilities were able to be updated because of the acquisition of a set of new information. Thus when a particular type of information is observed, a specific action is chosen, based on expected values. If expected opportunity losses are used as the payoffs, then the decision strategy for each type of information to minimize expected opportunity loss is referred to as the Bayes strategy.

The Bayes-strategy expected payoff is a weighted average of the expected opportunity losses. Each minimum expected opportunity loss under each type of information is weighted by the probability of observing that type of information. The value of information is the amount of opportunity loss decreased by the Bayes strategy over the prior minimum expected opportunity loss. The benefits derived by the Bayes strategy must be worth the cost of obtaining the necessary information.

Game Theory

Game theory concerns that aspect of decision analysis under conflict where the states of nature are under the control of an intelligent adversary. Probabilities associated with the competitor's choice of action will reflect his attempt to minimize the decision maker's payoff (or maximize the losses). Each payoff $cell_{i,j}$ in the decision matrix has a probability of happening equal to the product of the two marginal probabilities, P(opponent picks action i) and P(decision maker choses action j).

The value of the game to the decision maker is the sum of all cell values in the decision matrix, multiplied by their probability of occurring. The marginal probabilities assigned to the actions are found by assuming that the value of the game for the decision maker (or the opponent) must be the same no matter which action or strategy the opponent chooses.

KEY TERMS

Bayes strategy	A strategy utilizing Bayesian decisions on each type of an information set. When a particular type of information is observed, the action is chosen that gives the minimum expected opportunity loss.
Bayes' theorem	A theorem to derive a conditional probability of a state of nature, using the prior probability of the state of nature and newly acquired information.
Bayesian decision rule	The choosing of the action that has the maximum expected payoff.
Decision matrix	A matrix of payoffs resulting from the combination of a set of decision alternatives and a set of states of nature.
Expected payoff	The weighted value of an action in which each payoff possibility of the action is multiplied by the probability of the state of nature for that payoff and the products summed.
Game theory	A branch of decision analysis in which the states of nature are controlled by an intelligent adversary.
Maximax rule	The rule dictating that the decision maker choose the action that would result in the most maximum payoff possible.
Minimax rule	The rule dictating that the decision maker choose the action that has the minimum of the various actions' maximum losses.
Opportunity loss	The difference between the maximum gain (or minimum loss) and the smaller gain

	(larger loss) from each of the alternative actions for each state of nature.
Posterior probabilities	The conditional probability of a state of nature, given the result of an observation not used before. A revision or update of the marginal prior probability.
Prior probabilities	The probabilities of states of nature before any observations are made to gain new information about its occurrence.
States of nature	Factors affecting the outcome of a decision over which the decision maker has no control.
Subjective probability	Probabilities based on the knowledge, intuition, guess, or hunch of the decision maker. Their validity may be questionable, but they are often useful in providing needed information to reduce uncertainty.
Value of information	The reduction in the expected opportunity loss resulting in the use of Bayes strategy.
Value of the game	The long-run or average benefit for the decision maker for each decision.

FORMULA REVIEW

(14-1) Expected payoff for each course of action

$$\sum_{i=1}^{n} C_i P(S_i)$$

for i states of nature where

C_i = payoff for state of nature i

$P(S_i)$ = probability of state of nature i

n = number of possible states of nature

(14-2) Bayes' formula

$$P(S_i | I) = \frac{P(S_i \cap I)}{P(I)} = \frac{P(S_i) P(I|S_i)}{\sum_{i=1}^{n} P(S_i) P(I|S_i)}$$

where I = new observation (gained information)

ORGANIZED LEARNING QUIZ

1. If an experimenter fails to reject a true hypothesis, he has made a

a. correct decision.
   b. type I error.
   c. type II error.
   d. decision based on too small a sample size.

2. Decision making under certainty is a situation where

   a. several states of nature may exist.
   b. several courses of action may exist.
   c. the decision maker cannot select the most favorable consequence for certain.
   d. incomplete information makes the decision completely a random one.

3. The maximax rule guides the decision maker to

   a. avoid purchasing lottery tickets.
   b. attempt to maximize the maximum possible opportunity loss.
   c. take the most pessimistic attitude on the future state of nature.
   d. obtain the maximum possible payoff from the available courses of action.

4. The expected payoff is calculated by summing the

   a. products of payoffs and probabilities of states of nature.
   b. products of probabilities of payoffs and states of nature.
   c. products of courses of action and probabilities of payoffs.
   d. products of probabilities of courses of action and payoffs.

5. The cells in a decision matrix represent the

   a. probabilities associated with each course of action.
   b. payoffs expressed as gains, profits, losses, or costs.
   c. posterior probabilities of states of nature.
   d. results of Bayesian actions and a Bayesian strategy.

6. Subjective probabilities refer to probabilities

   a. derived after taking many observations of experiment outcomes.
   b. with values between -1 and +1.
   c. reflecting the beliefs of the decision maker.
   d. conditioned by the future state of nature.

7. Bayes' formula computes a(n)

   a. marginal probability of a course of action.
   b. joint probability of a payoff and a state of nature.
   c. conditional probability of a state of nature given additional information.
   d. independent observation for a random sample.

8. In this chapter the value of information is computed by measuring a

   a. change in results of a course of action.
   b. change in the states of nature probabilities.
   c. decrease in the maximum expected opportunity loss.
   d. decrease in the minimum expected opportunity loss.

9. Game theory assumes that

   a. the objective of an opponent is to split the payoffs obtained evenly.
   b. states of nature are the courses of action for an intelligent opponent.
   c. an adversary makes decisions irrationally.
   d. an opponent determines payoffs listed in the decision matrix.

10. To determine the value of a game when using mixed strategies,

    a. the assumption is made that the value remains unchanged regardless of the action taken by the opponent.
    b. one calculates the minimum expected opportunity loss.
    c. the assumption is made that the value depends on the state of nature alone.
    d. one calculates the maximax.

11. A decision is said to be made under _____ when there is no doubt as to the state of nature.

12. Probabilities used in a decision matrix are (always, sometimes, never) based on exact knowledge of the state of nature.

13. If the marginal probabilities of A and B are P(A) and P(B) and their values are .8 and .4, respectively, and the conditional probability of B, given A has already occurred, if $P(B|A)$ equals .2, then how is $P(A|B)$ calculated? _____

14. Opportunity losses are labeled as "opportunity" because the decision maker has a possible chance of maximum gain. A loss of that maximum only occurs if he or she _____.

15. The _____ rule minimizes the maximum possible _____ for each alternative action.

16. Explain why the Bayer' formula is said to compute conditional probabilities "in reverse."

    _____
    _____
    _____

17. Explain why, in computing action probabilities in game theory, the expected payoffs under each state of nature are set equal to each other.

    _____
    _____
    _____

18. Explain how to apply the Bayes' strategy procedure and its expected payoff.

    _____

19. Explain why a decision matrix can be useful when the payoffs are measured in losses as well as gains.

20. Under what conditions would one want to use the maximax rule?

ANSWERS TO ORGANIZED LEARNING QUIZ

1. a   2. b
3. d   4. a
5. b   6. c
7. c   8. d
9. b   10. a

11. certainty

12. never

13. $\dfrac{P(A)P(B|A)}{P(B)} = \dfrac{(.8)(.2)}{.4} = \underline{.4}$

14. chooses the wrong action

15. minimax, loss

16. Knowing a prior marginal probability of a state of nature and having additional information, it is possible to invert $P(I|S)$ to $P(S|I)$, called the posterior probability.

17. The best possible solution under mixed strategy is one where the expected payoff is the same regardless of the choice of the opponent.

18. Posterior probabilities are applied to states of nature; then, for each possible outcome of an observation, the best alternative action is identified. By taking a weighted average of the best action results, the expected payoff is obtained. Each action result is weighted by the probability of the observation outcome that generated that result.

19. Procedures used to find the best alternative course of action can be based on the objective of maximizing gains or minimiz-

ing losses. The results are equivalent.

20. The maximax rule chooses that course of action among all courses of action that has the highest return as a possible payoff. Highly optimistic, high-risk takers, or highly desperate decision makers will probably choose this course of action. The risk of failure is not high enough to deter the lure of the high payoff that is possible, no matter how slight its possibility might be.

SAMPLE EXERCISE

Given the following decision matrix of payoff gains:

	A_1	A_2	A_3
S_1	40	50	40
S_2	40	30	40
S_3	60	20	40
S_4	10	70	40

1. Which alternative is the maximax decision?

2. Construct the matrix of opportunity losses.

	A_1	A_2	A_3
S_1			
S_2			
S_3			
S_4			

Assume: $P(S_1) = .1$; $P(S_2) = .3$; $P(S_3) = .4$; $P(S_4) = .2$
(priors)

3. What is the expected gain from each alternative decision? Which is optimal?

4. What is the expected opportunity loss from each alternative decision? Which is optimal? Compare your answer with problem 3. Also assume: Additional information is available with two outcomes with probabilities $P(I_1) = .6$ and $P(I_2) = .4$. Moreover, $P(I_1|S_1) = 1.0$ $P(I_1|S_2) = 1/3$ $P(I_1|S_3) = .75$
$P(I_1|S_4) = .5$ $P(I_2|S_1) = 0$ $P(I_2|S_2) = 2/3$
$P(I_2|S_3) = .25$ $P(I_2\,S_4) = .5$

5. What is the Bayes action for each information outcome (Bayes'

strategy)?

6. What is the value of the information?

ANSWERS TO SAMPLE EXERCISE

1. A_2 to hope for the payoff gain of 70.

2.
	A_1	A_2	A_3
S_1	10	0	10
S_2	0	10	0
S_3	0	40	20
S_4	60	0	30

3.
	A_1	A_2	A_3	$P(S_i)$
S_1	40	50	40	.1
S_2	40	30	40	.3
S_3	60	20	40	.4
S_4	10	70	40	.2

Expected gain: 42* 36 40   Therefore A_1 is optimal.

4.
	A_1	A_2	A_3	$P(S_i)$
S_1	10	0	10	.1
S_2	0	10	0	.3
S_3	0	40	20	.4
S_4	60	0	30	.2

Expected opportunity loss: 13* 19 15   Therefore A_1 is optimal (solution the same).

5. Using the table of opportunity losses for ease of computation,

177

	A_1	A_2	A_3	$P(S_i)$	$P(S_i\|I_1)$	$P(S_i\|I_2)$
S_1	10	0	10	.1	1/6	0
S_2	0	10	0	.3	1/6	1/2
S_3	0	40	20	.4	1/2	1/4
S_4	60	0	30	.2	1/6	1/4
I_1:	$11\frac{2}{3}$*	$21\frac{2}{3}$	$16\frac{2}{3}$			
I_2:	15	15	$12\frac{1}{2}$*			

Therefore Bayes strategy is: If I_1 occurs, then A_1 is chosen.
If I_2 occurs, then A_3 is chosen.

6. Since $P(I_1) = .6$ and $P(I_2) = .4$, the expected payoff of Bayes strategy is an expected opportunity loss of

$$(11\tfrac{2}{3})(.6) + (12\tfrac{1}{2})(.4) = 7 + 5 = \underline{12}.$$

The value of information is the difference of the answer in problem 4 and the answer in problem 6:

$$13 - 12 = \underline{1}.$$

CASE PROBLEM

Saturday Market

Catherine Bruening and Bill Blodgett have each had the same hobby over the past few years--making miniature dolls and dollhouses. They have used them in the past as presents for their relatives and friends. The quality of their work has improved to a point where they are told they should put a price on them and offer them to the general public.

Every Saturday during the summer a section of the downtown area under a bridge approach is blocked off by the city and small square spaces made available to individuals wishing to construct booths. Ethnic foods, jewelry, clothing, and small gifts are on display and are for sale to the public. Bill and Cathy have decided to rent a space, construct a portable, knockdown variety of a booth, and exhibit their works. Tags showing the price of each article are made and attached to each piece. They are not sure whether many sales will result, but the fun they expect to have will more than compensate for any disappointment over a lack of sales.

If Bill and Cathy knew exactly what the demand for their product would be, they would know exactly how many and which of their items to bring for display. The size of the dollhouses and the small display space have limited the number; however, they feel that a large proportion of their stock can be shown.

As long as an effort is being made in this direction, Bill and Cathy have agreed they should maximize their net income from their operation. They have narrowed their alternative displays to three possibilities, each with a probable sales potential, depending on the weather, number of people at the market, their propensity for purchasing their type of wares, and so on. They feel that three outcomes, over which they have little or no control, are possible. They have classified them as excellent conditions, fair conditions, or poor conditions. They also feel that they have an idea as to the probability of each condition occurring, but they also believe that they would be more confident of these probabilities by talking to some of the current booth managers.

At this point they are at a loss as how to put the information they have or might obtain to use to help them decide which alternative display should be used. Furthermore, they are not sure whether gathering more information by talking to other booth managers will be worthwhile, as measured by additional profits gained.

After discussing their dilemma with their mutual friend, Gary, he says that there are procedures available to help them make a choice. He has offered to spend some time with them, do some calculations based on decision theory, and give them an answer as to which display to use.

1. What are the steps necessary for a decision?

2. What probabilities are necessary?

15

SIMPLE REGRESSION AND CORRELATION

CHAPTER LEARNING OBJECTIVES

On completing this chapter the student should be able to

- Explain the concepts related to regression analysis and relate them to the concepts of correlation analysis.

- Describe the graphical meaning of the parameters in a regression equation.

- Interpret a scatter diagram as whether it is conducive to linear analysis.

- Define the criterion of line fit called the least squares.

- Explain the rationale of using the regression line in statistical inference.

- Explain the meaning and describe the usefulness of the coefficient of correlation and the coefficient of determination in correlation analysis.

- Do the mathematical calculations necessary to determine the regression equation and the correlation coefficient.

- Determine which of two variables should be considered the dependent and independent variable in regression analysis.

SYNOPSIS

The existence and strength of a relationship between two variables are investigated here. The methods studied are labeled regression and correlation techniques.

Related Concepts

In the regression method a limited model involving a linear relationship between two variables is chosen. A simple linear regression equation involves two variables, an independent or predict*or* variable (X) and a dependent or predict*ed* variable (Y). The linear (straight-line) equation involving the variables can be expressed mathematically as

$$Y = a + bX$$

where a and b are the parameters of the equation and their values determine the relationship that may exist between the variables. Finding the values of the parameters a and b is one of the purposes of regression analysis.

Care must be taken when interpreting an existing relationship between two variables in that cause and effect are not necessarily implied. In the equation above a value given for X will result in a unique value for Y (assuming a and b have been predetermined). This result is functionally or consequentially determined and not necessarily caused by the given value of X. There may exist a third variable as a cause for the relationship given between X and Y. Notice also that the inverse function obtained by solving X as a linear function of Y does not imply that values of Y "cause" particular values for X. The study of regression techniques centers on the results of relationships without being concerned with their causes. The measure of closeness of the relationship between variables is determined by correlation analysis. Correlation analysis can be done without developing a regression equation; however, since the computations for both are similar, they are usually done at the same time.

Linear Regression

When an investigator is presented paired data, usually the first step in an analysis is to plot a scatter diagram. Afterward a least squares method is used to derive a regression equation for the sample data. Inferences about the true regression equation parameters can then be made and, finally, dependent Y values estimated on the basis of given independent X values.

The Scatter Diagram. A plot of the data with values of X on the horizontal axis and values of Y on the vertical axis often gives a visual hint of the existence of a linear relationship. A freehand line through the points "averaging" the distances that points have above and below the line is one way to provide a rough estimate of the regression line. Such a method gives no insurance that the line drawn has done the best job of averaging.

The Method of Least Squares. The method used to determine mathematically the best averaging line is an extension of the concept of averaging when only one variable is present. One attribute of the mean of a set of values of one variable is that the sum of the squares of the distance each observation is from the mean is at a minimum. Any other value but the mean will give a larger sum of squared deviations.

For two variables, the concept is the same except that the mean will not be a point but a line, the desired regression line.

To determine the values of a and b for the regression line, we find the particular values of a and b that make the sum of the squared distance that each point is off the line the smallest. In higher mathematics it is shown that minimizing the sum of squares results in two simultaneous equations in variables a and b. They are called the *normal equations of regression* and their solution gives values for both a and b.

The plotted regression equation gives predicted values of Y when predictor values of X are given. These values can either be read from the plot or computed from the linear equation. The computed values of Y are labeled Y_c.

Inference about the Regression Coefficient. Since the values of a and b were determined from sample data, the true parameters of the regression line can be labeled ∝ and β, where ∝ is the true Y intercept and β is the true slope. The sample regression line is an estimate of the true conditional mean of Y, given X.

Computation of the Standard Deviation of Y. For any X, the observed variable Y varies in a random manner with a mean of $\mu_{Y|X} = \alpha + \beta X$ and standard deviation $\sigma_{Y|X}$. Assuming a normal distribution, the mean is estimated with Y_c and the standard deviation is estimated with $s_{Y|X}$. Two degrees of freedom are lost because of estimating ∝ and β. $s_{Y|X}$ is called the estimated conditional standard deviation of Y from Y_c.

Hypothesis Testing and Interval Estimation about β. The sample statistic b is used to estimate β. To make inferences about β, something must be known about the sampling distribution of b. The assumption is made that b is normally distributed with a mean β and a standard error of σ_b estimated as s_b. The null hypothesis that β is not significantly different from zero can be tested and confidence intervals for β found.

Estimation of the Conditional Mean $\mu_{Y|X}$. What is desired in this section is an expression for the confidence interval for the expected value of Y, given X. The sampling distribution of Y_c is normal with mean $\mu_{Y|X}$ and standard error of σ_{Y_c} estimated as s_{Y_c}, which is a function of X and is minimized when $X = \bar{X}$. The confidence interval for $\mu_{Y|X}$ also becomes wider as X moves farther from $\bar{X}$.

Estimation of the Actual Y Value. If the actual value of Y, given X, is denoted as Y_a, then the standard deviation is σ_{Y_a} and is estimated by s_{Y_a}.

Simple Correlation

An accurate prediction of Y, given X, is only possible if there is a strong relationship between X and Y and if that relationship is given in the least squares equation. A weak relationship will result in a weak model. The measure of the strength of correlation of two variables is the correlation coefficient r for a sample and ρ (rho) for the population.

The Coefficient of Correlation. If the regression line can eliminate a large part of the variation of the Y's, then a better estimate can be obtained. Variation of the Y's can be partitioned into that variation eliminated by the regression line and the variation that remains in spite of the regression line. Therefore the proportion of the total that is explained or eliminated plus the proportion remaining sums to one.

The proportion eliminated or explained by the regression equation is labeled r^2 and is called the coefficient of determination. Its square root, r, is the coefficient of correlation.

$$r = \sqrt{1 - \frac{\text{unexplained variation}}{\text{total variation}}} = \sqrt{1 - \frac{\Sigma (Y - Y_c)^2}{\Sigma (Y - \bar{Y})^2}}$$

The computational form, using the computation table in the Sample Exercise below, is

$$r = \sqrt{\frac{(b)(\text{covariation})}{\text{total variation}}} = \sqrt{\frac{(b)(\Sigma XY - \bar{X} \Sigma Y)}{\Sigma Y^2 - \bar{Y} \Sigma Y}}$$

For r values close to zero, very little relationship exists. For r values close to +1, strong positive correlation exists and X and Y move in the same direction. For values of r close to -1, strong negative correlation exists and the variables X and Y move in opposite directions; i.e., rising values of X are accompanied by falling values of Y.

Test of Hypothesis. To test the null hypothesis that rho is not significantly different from zero, the test statistic

$$T = r\sqrt{\frac{n - 2}{1 - r^2}}$$

is used if an assumption can be made of normality of the two variables. If not, then the nonparametric rank correlation test is a convenient substitute. Rejection occurs if

$$T \geq t_{n-2, \alpha/2} \quad \text{or} \quad T \leq -t_{n-2, \alpha/2}$$

KEY TERMS

Coefficient of determination	The proportion of variance in the dependent variable that is explained by the regression line. The square of r.
Correlation analysis	A procedure to determine how closely a linear relationship exists between variables included in a regression equation.
Correlation coefficient	A measure of the strength of correlation based on a sample of matched observations.

	Denoted as r for the sample and ρ (rho) for the population.
Dependent variable	The variable to be predicted or estimated based on the value of the independent variable.
Independent variable	The variable that provides the basis for the estimation of the value of the dependent variable. The independent variable is called the predictor variable.
Least squares criterion	The assurance that the line used to fit the sample data minimizes the sums of squares of the vertical distances from the points to the line.
Linear regression	The relationship between variables that can be displayed as a straight line.
Regression analysis	Simple regression procedures derive an equation to fit a collection of sample data involving a dependent variable and an independent variable. Multiple regression procedures derive an equation to fit a collection of sample data involving one dependent variable and two or more independent variables.
Regression coefficients	The parameters of a regression equation, the value of which determines the location and slope of the regression line.
Scatter diagram	A visual plot of the sample data points that possibly suggests the kind of relationship existing between variables.
Univariate analysis	Problems concerned and involving only one variable.
Variation	The sum of the squared distances of a set of observations from their mean. This is also called the total variation when applied to the dependent variable in regression analysis. The total variation is the sum of the variation explained by the regression line and the variation not explained by the regression line.

FORMULA REVIEW

(15-1) Regression line $Y = a + bX$

where $b = \dfrac{\text{covariation}}{\text{variation of } X} = \dfrac{\Sigma XY - \bar{X} \Sigma Y}{\Sigma X^2 - \bar{X} \Sigma X}$

and $a = \bar{Y} - b\bar{X}$

(15-8) Conditional standard error of Y, given X

$$s_{Y|X} = \sqrt{\frac{\text{unexplained variation}}{n-2}}$$

$$= \sqrt{\frac{\Sigma (Y - Y_c)^2}{n-2}}$$

$$= \sqrt{\frac{\Sigma Y^2 - a\Sigma Y - b\Sigma XY}{n-2}}$$

(15-9) Sample standard error of b

$$s_b = \frac{s_{Y|X}}{\sqrt{\Sigma (X - \bar{X})^2}} = \frac{s_{Y|X}}{\sqrt{\Sigma X^2 - \bar{X}\Sigma X}}$$

(15-10) The b test statistic

$$T = \frac{b - \beta}{s_b}$$

(15-12) Sample standard deviation of the conditional mean Y_c.

$$s_{Y_c} = s_{Y|X} \sqrt{\frac{1}{n} + \frac{(X - \bar{X})^2}{\Sigma X^2 - \bar{X}\Sigma X}}$$

(minimum when $X = \bar{X}$)

(15-15) Sample standard deviation of actual value of Y_a.

$$s_{Y_a} = s_{Y|X} \sqrt{1 + \frac{1}{n} + \frac{(X - \bar{X})^2}{\Sigma X^2 - \bar{X}\Sigma X}}$$

(minimum when $X = \bar{X}$)

(15-17) Total variation = explained var. + unexplained var.

$$\sum_{i=1}^{n}(Y_i - \bar{Y})^2 = \sum_{i=1}^{n}(Y_{c_i} - \bar{Y})^2 + \sum_{i=1}^{n}(Y_i - Y_{c_i})^2$$

(variation in Y) = (b)(covariation) + (variation in Y) − (b)(covariation)

(15-19) Coefficient of determination

$$r^2 = 1 - \frac{\text{unexplained variation}}{\text{total variation}}$$

$$= 1 - \frac{\Sigma(Y - Y_c)^2}{\Sigma(Y - \bar{Y})^2} = \frac{\text{explained variation}}{\text{total variation}}$$

$$= \frac{(b)(\text{covariation})}{\text{variation of Y}}$$

(15-22) Correlation coefficient test statistic T

$$T = r\sqrt{\frac{n-2}{1-r^2}}$$

ORGANIZED LEARNING QUIZ

1. Univariate analysis refers to problems involving

    a. only one observation.
    b. only one variable.
    c. only one statistic.
    d. only one pair of variables.

2. Regression techniques involving one dependent variable and two or more independent variables are called

    a. simple regression analysis.
    b. correlation analysis.
    c. straight-line analysis.
    d. multiple regression analysis.

3. The criterion used to determine the best line to draw through a set of observation points is that

    a. there must be as many points above the line as below the line.
    b. the sum of the absolute values of the vertical distance each point is off the line must be at a minimum.
    c. the sum of the squared values of the vertical distance each point is off the line must be at a minimum.
    d. the line must go through as many points as possible.

4. In the regression equation the value of the Y intercept and the value of the slope are known as

    a. the regression coefficients.
    b. the coefficients of determination.
    c. the regression variables.
    d. the independent and dependent variables.

5. The regression equation is stated as

    a. $\bar{Y} = a + b\bar{X}$.
    b. $Y = aX - b$.
    c. $Y = X + ab$.
    d. $Y = a + bX$

6. The regression equation defines a line referred to as a conditional mean because

    a. the computed value of Y is what is expected, given the value of X used.
    b. the computed value of Y is correct only if the line slopes upward.
    c. the computed value of Y is expected, given a large sample size only.

d. the computed value of Y must be smaller than values of X used.

7. It is more difficult to estimate an actual value of Y than an expected value for Y because

   a. the expected Y value is constant for all samples.
   b. the actual Y values have greater variations.
   c. no formulas are available to compute a measure of an estimate of the variability of actual Y values.
   d. the standard deviation of the actual value for Y is smaller than for the expected value.

8. Variation of the data points about the horizontal line through $\bar{Y}$ compared to the variation about the regression line is

   a. smaller.
   b. equal.
   c. smaller by 1/n times.
   d. larger.

9. The coefficient of determination

   a. can be negative.
   b. is the proportion of total variation left unexplained by the regression line.
   c. is a measure of the variation of error.
   d. is always greater than the correlation coefficient.

10. A negative value calculated for the correlation coefficient means

    a. a calculation error--r^2 is always positive.
    b. the value of a in the regression equation is negative.
    c. the relationship is negative; an increase in X is accompanied by a decrease in Y.

11. The _____ variable is the one to be estimated or predicted and the _____ variable is the one providing the basis for estimation.

12. A causal relationship implies that the dependent variable is the (cause, effect) and the independent variable is the (cause, effect).

13. Each pair of observations in a data sample is represented by a _____ on the scatter diagram.

14. The regression coefficient giving the slope of the line is labeled _____, while the other coefficient tells _____.

15. The standard deviation of the calculated value of Y is smallest at the _____ value of X.

16. Why is it important to plot a scatter diagram of the data in a simple linear regression model?

17. Explain the difference between the terms "functional" relationship and "causal" relationship. Which is implied in regression analysis?

18. What is meant by the coefficient of determination?

19. What is meant by "least squares" in a regression model?

20. Distinguish between dependent and independent variables in a regression model.

ANSWERS TO ORGANIZED LEARNING QUIZ

1. b
2. d
3. c
4. a
5. d
6. a
7. b
8. d
9. d
10. c

11. dependent, independent

12. effect, cause

13. point

14. b, where the line crosses the Y axis

15. mean

16. A plot of the data might suggest some relationship other than a linear one, in which case linear regression analysis would be inappropriate.

17. If a functional relationship exists between two variables, then it can be represented by a formula and the relationship itself shown, not the cause of the relationship. No attempt is made in regression analysis to determine cause and effect, if it exists, in a relationship.

18. The coefficient of determination measures the proportion of the total variability of the dependent variable that is reduced or explained by using the regression line as an estimating device instead of its mean. It is a measure of how good the predicting model is, since the quality of any estimator is enhanced if the variation in the predicted variable is reduced.

19. Least squares refers to the line that minimizes the sum of squares of the vertical distance that each point is off the line. The assumptions in linear regression analysis are (a) there is a probability distribution of Y for each value of X and (b) the means of these probability distributions fall perfectly on a line. The least squares criterion in determining that line is built on the property that the least squares estimators, a and b, are unbiased and are excellent estimators of the population parameters α and β.

20. The variable used to predict the variable of interest is called the independent variable and the variable being predicted is called the dependent variable.

SAMPLE EXERCISE

It is desired to find if there is a relationship between gasoline sales in a county and the spendable income for the county. A sample of ten counties produced the figures shown in the table.

1. Find the regression coefficients and the regression line.

2. Calculate the total variation, explained variation, and the unexplained variation.

3. Calculate the value for r^2 and r.

4. Test the null hypothesis that rho is not significantly different from zero (alpha = .05).

County	Income ($10,000,000)	Sales ($100,000)
1	16	40
2	24	50
3	32	68
4	15	36
5	20	45
6	12	27
7	18	42
8	14	36
9	10	29
10	29	67

Following the tabular computation approach utilized in previous chapters, the data are organized in the fashion shown below.

Pair No.	X	Y	X^2	XY	Y^2
1	16	40	256	640	1600
2	24	50	576	1200	2500
3	32	68	1024	2176	4624
4	15	36	225	540	1296
5	20	45	400	900	2025
6	12	27	144	324	729
7	18	42	324	756	1764
8	14	36	196	504	1296
9	10	29	100	290	841
10	29	67	841	1943	4489
Totals	190 Ⓐ	440 Ⓑ	4086 Ⓒ	9273 Ⓓ	21164 Ⓔ
Means and corrections	19 Ⓕ	44 Ⓖ	3610 Ⓗ	8360 Ⓘ	19360 Ⓙ
Variations and covariation			476 Ⓚ	913 Ⓛ	1804 Ⓜ

Explanation of Table and Calculations

Ⓐ $= \sum_{i=1}^{n} X_i$
sum of X's

Ⓑ $= \sum_{i=1}^{n} Y_i$
sum of Y's

Ⓒ $= \sum_{i=1}^{n} X_i^2$
sum of X^2's

Ⓓ $= \sum_{i=1}^{n} X_i Y_i$
sum of products

Ⓔ $= \sum_{i=1}^{n} Y_i^2$
sum of Y^2's

Ⓕ $= \bar{X} = \frac{Ⓐ}{n}$
mean of X

$$\text{Ⓖ} = \bar{Y} = \frac{\text{Ⓑ}}{n} \qquad\qquad \text{Ⓗ} = \bar{X}\sum_{i=1}^{n}X_i = \text{Ⓕ} * \text{Ⓐ}$$

mean of Y

$$\text{Ⓘ} = \bar{X}\sum_{i=1}^{n}Y_i = \text{Ⓕ} * \text{Ⓑ} \qquad\qquad \text{Ⓙ} = \bar{Y}\sum_{i=1}^{n}Y_i = \text{Ⓖ} * \text{Ⓑ}$$

$$\text{Ⓚ} = \sum_{i=1}^{n}(X_i - \bar{X})^2 = \text{Ⓒ} - \text{Ⓗ} \qquad \text{Ⓛ} = \sum_{i=1}^{n}(X_i-\bar{X})(Y_i-\bar{Y}) = \text{Ⓓ} - \text{Ⓘ}$$

variation of X covariation

$$\text{Ⓜ} = \sum_{i=1}^{n}(Y_i - \bar{Y})^2 = \text{Ⓔ} - \text{Ⓙ}$$

variation of Y (total variation)

Slope b is computed by Ⓛ/Ⓚ = 913/476 = <u>1.918067</u> Ⓝ.

Intercept a is Ⓖ - Ⓝ * Ⓕ = 44 - (1.918067)(19) = <u>7.556723</u>.

Explained variation is Ⓝ * Ⓛ = (1.918067)(913) = <u>1751.195</u> Ⓟ.

Unexplained variation is Ⓜ - Ⓟ = 1804 - 1751.195 = <u>52.80463</u>.

Total variation is Ⓜ = <u>1804</u>.

r^2 = Ⓟ/Ⓜ = 1751.195/1804 = <u>.9707291</u> Ⓠ.

r = $\sqrt{Ⓠ}$ = $\sqrt{.9707291}$ = <u>.9852559</u>

$$T = r\sqrt{\frac{n-2}{1-r^2}} = (.9852559)\sqrt{\frac{8}{1 - .9707291}} = \underline{16.28832}$$

T is considerably greater than the t value of 2.306; therefore the null hypothesis is rejected.

CASE PROBLEM

In order to qualify for admission to graduate school in a large urban university, a prospective student must complete several application requirements. One is to have an acceptable score on an aptitude test given nationwide. The purpose of the test is to measure whether a person has the proper background and basic tools necessary for study at the graduate level. The graduate school of business in the university has used the nation's median score as a minimum for admission for several years and Dr. Donald Moore, the assistant dean of the school, has always had an uneasy feeling about this requirement.

He realizes that a person's score is not an absolute measure but may be a reflection of several factors not associated with what the test is designed to measure. One is how the test taker physically feels on the day of the test. Because of the score variability for any individual, errors of two kinds are committed. Some adequate students are denied admission and some inadequate students are passed.

Dr. Moore decides to make a study of past students to discover whether there is any relationship between the test score and student performance after admittance. This factor has become even more important at this particular time because the faculty graduate committee has recommended that the cutoff point be raised to, in the committee's words, "further limit the enrollment to persons more likely to finish their graduate studies."

As a first test design, he has decided to test the null hypothesis of no significant relationship between entrance scores and dropout rate among business students.

The data below were gathered and classified to be more compatible with simple regression and correlation analysis. The second column lists the number of students entering the program within each entrance score classification. The last column is the percentage of those students who had not finished by the end of their fourth year. This time frame was chosen because the data show that over 95% of graduates from the school had finished within four years.

Score (percentile)	Number of Students Enrolled	Number of Students not Finishing	Percent
50 - 54	96	7	7.29
55 - 59	141	9	6.38
60 - 64	129	5	3.88
65 - 69	105	8	7.62
70 - 74	153	4	2.61
75 - 79	144	5	3.47
80 - 84	84	4	4.76
85 - 89	72	2	2.78
90 - 94	36	2	5.56
95 - 99	18	0	0

COMPUTER PROGRAM

The BASIC program following has a limit of 50 pairs of observations. Sample runs are given for the Sample Exercise and for the Case Problem.

```
10 REM*********************************************************
20 REM
30 REM      TITLE:       LINEARPO
40 REM      LANGAUGE:    BASIC
50 REM      DESCRIPTIONS:
60 REM           THIS PROGRAM IS FOR ONE INDEPENDENT VARIABLE
70 REM           REGRESSION.
80 REM      INSTRUCTIONS:
90 REM           ENTER DATA AS REQUESTED BY THE PROGRAM.
100 REM     PROGRAMMER:
110 REM          GROVER WM. RODICH, PORTLAND STATE UNIV.
120 REM
130 REM*********************************************************
140 REM
150 KEY OFF
160 DIM X(50),Y(50)
170 CLS
```

```
180 PRINT"SIMPLE LINEAR REGRESSION AND CORRELATION PROGRAM"
190 PRINT
200 PRINT "      INPUT 0 TO EXIT PROGRAM"
210 PRINT "      INPUT 1 TO CONTINUE"
220 PRINT:PRINT "      YOUR CHOICE";:INPUT G
230 IF G=0 THEN 930
240 IF G<>1 THEN 200
250 CLS
260 PRINT "PLEASE INPUT THE  NUMBER OF OBSERVATION PAIRS"
270 PRINT "N= ";
280 INPUT N
290 PRINT "INPUT THE DATA PAIRS SEPARATED BY A COMMA, ONE DATA"
300 PRINT "PAIR AT A TIME FOLLOWED BY A RETURN."
310 PRINT "TYPE X FIRST, THEN Y"
320 FOR I=1 TO N
330 INPUT X(I),Y(I)
340 NEXT I
350 PRINT
360 PRINT TAB(5); "PAIR #";TAB(20);"X";TAB(30);"Y";TAB(39);"X SQ";TAB(49);
370 PRINT "XY";TAB(59);"Y SQ"
380 PRINT
390 FOR I=1 TO N
400 PRINT TAB(3);I;TAB(17);X(I);TAB(27);Y(I);TAB(37);X(I)^2;
410 PRINT TAB(47);X(I)*Y(I);TAB(57);Y(I)^2
420 NEXT I
430 PRINT "----------------------------------------------------------------"
440 PRINT
450 A=0
460 B=0
470 C=0
480 D=0
490 E=0
500 FOR I=1 TO N
510   A=A+X(I)
520   B=B+Y(I)
530   C=C+X(I)^2
540   D=D+X(I)*Y(I)
550   E=E+Y(I)^2
560 NEXT I
570 PRINT "TOTALS";TAB(17);A;TAB(27);B;TAB(37);C;TAB(47);D;TAB(57);E
580 A1=A/N
590 B1=B/N
600 C1=A1*A
610 D1=A1*B
620 E1=B1*B
630 PRINT
640 PRINT "MEANS AND";TAB(17);A1;TAB(27);B1;TAB(37);C1;TAB(47);
650 PRINT D1;TAB(57);E1
660 PRINT "CORRECTION";TAB(37);"------------------------------"
670 PRINT
680 C2=C-C1
690 D2=D-D1
700 E2=E-E1
710 PRINT "VARIATIONS AND COVARIATION";TAB(37);C2;TAB(47);
720 PRINT D2;TAB(57);E2
730 PRINT
740 PRINT
750 LOCATE 24,1:PRINT "TO COMPLETE OUTPUT PRESS ANY KEY";:A$=INPUT$(1)
760 PRINT:PRINT
770 PRINT "SLOPE B= ";D2/C2;"INTERCEPT A= ";B1-(D2/C2)*A1
780 PRINT
790 PRINT "EXPLAINED VARIATION";TAB(25);(D2/C2)*D2
800 PRINT "UNEXPLAINED VARIATION ";TAB(25);E2-(D2*D2/C2)
810 PRINT "TOTAL VARIATION";TAB(25);E2
820 PRINT
830 PRINT "COEFFICIENT OF DETERMINATION, R-SQUARE EQUALS";D2*D2/C2/E2
840 PRINT
850 PRINT "COEFFICIENT OF CORRELATION, R EQUALS";(D2*D2/C2/E2)^.5
860 PRINT
870 PRINT "CORRELATION COEFFICIENT T STATISTIC EQUALS";
880 A3=(D2*D2/C2/E2)^.5
890 PRINT A3*((N-2)/(1-A3*A3))^.5
900 PRINT
910 LOCATE 24,1:PRINT "TO CONTINUE PRESS ANY KEY";:A$=INPUT$(1)
920 GOTO 170
930 KEY ON:END
```

SIMPLE LINEAR REGRESSION AND CORRELATION PROGRAM

```
    INPUT 0 TO EXIT PROGRAM
    INPUT 1 TO CONTINUE

    YOUR CHOICE? 1
```

PLEASE INPUT THE NUMBER OF OBSERVATION PAIRS
N= ? 10
INPUT THE DATA PAIRS SEPARATED BY A COMMA, ONE DATA
PAIR AT A TIME FOLLOWED BY A RETURN.
TYPE X FIRST, THEN Y
? 16,40
? 24,50
? 32,68
? 15,36
? 20,45
? 12,27
? 18,42
? 14,36
? 10,29
? 29,67

PAIR #	X	Y	X SQ	XY	Y SQ
1	16	40	256	640	1600
2	24	50	576	1200	2500
3	32	68	1024	2176	4624
4	15	36	225	540	1296
5	20	45	400	900	2025
6	12	27	144	324	729
7	18	42	324	756	1764
8	14	36	196	504	1296
9	10	29	100	290	841
10	29	67	841	1943	4489
TOTALS	190	440	4086	9273	21164
MEANS AND CORRECTION	19	44	3610	8360	19360
VARIATIONS AND COVARIATION			476	913	1804

TO COMPLETE OUTPUT PRESS ANY KEY

TOTALS	190	440	4086	9273	21164
MEANS AND CORRECTION	19	44	3610	8360	19360
VARIATIONS AND COVARIATION			476	913	1804

TO COMPLETE OUTPUT PRESS ANY KEY
SLOPE B= 1.918067 INTERCEPT A= 7.556725

EXPLAINED VARIATION   1751.195
UNEXPLAINED VARIATION   52.80469
TOTAL VARIATION   1804

COEFFICIENT OF DETERMINATION, R-SQUARE EQUALS .9707291

COEFFICIENT OF CORRELATION, R EQUALS .9852558

CORRELATION COEFFICIENT T STATISTIC EQUALS 16.2883

TO CONTINUE PRESS ANY KEY

SIMPLE LINEAR REGRESSION AND CORRELATION PROGRAM

```
    INPUT 0 TO EXIT PROGRAM
    INPUT 1 TO CONTINUE

    YOUR CHOICE? 1
```

```
PLEASE INPUT THE  NUMBER OF OBSERVATION PAIRS
N= ? 10
INPUT THE DATA PAIRS SEPARATED BY A COMMA, ONE DATA
PAIR AT A TIME FOLLOWED BY A RETURN.
TYPE X FIRST, THEN Y
? 52.5,7.29
? 57.5,6.38
? 62.5,3.88
? 67.5,7.62
? 72.5,2.61
? 77.5,3.47
? 82.5,4.76
? 87.5,2.78
? 92.5,5.56
? 97.5,0
```

PAIR #	X	Y	X SQ	XY	Y SQ
1	52.5	7.29	2756.25	382.725	53.1441
2	57.5	6.38	3306.25	366.85	40.7044
3	62.5	3.88	3906.25	242.5	15.0544
4	67.5	7.62	4556.25	514.35	58.0644
5	72.5	2.61	5256.25	189.225	6.8121
6	77.5	3.47	6006.25	268.925	12.0409
7	82.5	4.76	6806.25	392.7	22.6576
8	87.5	2.78	7656.25	243.25	7.7284
9	92.5	5.56	8556.25	514.3	30.9136
10	97.5	0	9506.25	0	0
TOTALS	750	44.35	58312.5	3114.825	247.1199
MEANS AND CORRECTION	75	4.435	56250	3326.25	196.6923
VARIATIONS AND COVARIATION			2062.5	-211.4251	50.42763

TO COMPLETE OUTPUT PRESS ANY KEY

TOTALS	750	44.35	58312.5	3114.825	247.1199
MEANS AND CORRECTION	75	4.435	56250	3326.25	196.6923
VARIATIONS AND COVARIATION			2062.5	-211.4251	50.42763

TO COMPLETE OUTPUT PRESS ANY KEY

SLOPE B= -.1025091 INTERCEPT A= 12.12318

EXPLAINED VARIATION   21.67299
UNEXPLAINED VARIATION   28.75463
TOTAL VARIATION   50.42763

COEFFICIENT OF DETERMINATION, R-SQUARE EQUALS .4297842

COEFFICIENT OF CORRELATION, R EQUALS .6555792

CORRELATION COEFFICIENT T STATISTIC EQUALS 2.45556

TO CONTINUE PRESS ANY KEY

SIMPLE LINEAR REGRESSION AND CORRELATION PROGRAM

```
    INPUT 0 TO EXIT PROGRAM
    INPUT 1 TO CONTINUE

    YOUR CHOICE? 0
OK
```

16

MULTIPLE REGRESSION AND CORRELATION

CHAPTER LEARNING OBJECTIVES

On completing this chapter the student should be able to

- Determine the coefficients of a multiple linear regression equation.

- Recognize a condition of complication in multiple regression analysis called collinearity.

- List the assumptions that form a basis for multiple regression analysis.

- Explain why there might be a difference between simple correlation coefficients and partial correlation coefficients.

- Explain the meaning of the coefficient of multiple determination.

- Calculate confidence intervals for the regression coefficients.

- Conduct tests of significance for the regression coefficients and the partial correlation coefficients.

SYNOPSIS

In a regression problem there is only one dependent variable, but there can be any number of independent variables. In the last chapter we investigated simple linear regression in which there was only one independent variable and the estimates of the dependent variable fell along a regression line. If there are two independent variables, the estimates of the dependent variable fall along a regression plane. Three or more independent variables generate estimates of the dependent variable that cannot be visualized but can be mathematically described as falling on a hyperplane.

The computations rapidly increase in complexity as the number of variables increase. Since the problems faced with multiple linear regression with more than one independent variable are merely an extension of the problems studied in simple linear regression, this chapter limits the discussion to problems involving only two independent variables and one dependent variable. First, an estimating equation will be found for Y where there is a single Y intercept (a) and a partial regression coefficient for each of the independent variables (b_1 and b_2). A measure comparable to the conditional standard deviation of Y_c is developed; also, a measure comparable to r^2 is found that measures the proportion of the total variation of the dependent variable explained by its relation to the two independent variables.

The Regression Equation

Using the method of least squares, a set of three simultaneous equations--the normal equations--in variables a, b_1, and b_2 can be found. Their solution generates the coefficients of the regression equation, expressed as

$$Y_c = a + b_1 X_1 + b_2 X_2 .$$

The effect of each independent variable on the dependent variable is linear, as shown by the preceding equation; however, a complication called collinearity or multicollinearity exists if the independent variables are highly correlated among themselves. If two independent variables have a high simple correlation coefficient, their partial regression coefficients might not be significantly different from zero when used together; but when used alone, their coefficients may be significantly different from zero. Tests to determine which variables are best to include in the final estimation equations require a large computational effort and should be done by computer.

Measuring the Error of Estimation

When two independent variables are used in a regression equation, the conditional standard error of the estimate is labeled $s_{Y \cdot 12}$

Multiple Correlation

The best measure of the strength of the relationship between the dependent variable and the multiple independent variables is the coefficient of multiple determination. This value measures the proportion of the total variation of the dependent variable that can be explained by relating it to the independent variables. Using the notation developed above, the coefficient is labeled $R^2_{Y \cdot 12}$

The symbol R^2 is used to distinguish this coefficient from the r^2 used in simple linear regression. The numerator of R^2 is in two parts--the variation explained by the first variable plus the variation explained by the second variable. You may remember that the numerator of r^2 contained only the variation explained by the single independent variable.

Partial Correlation

An investigator may wish to find a measure of the relationship of the dependent variable with any one of the independent variables. He can do so with the partial coefficient of correlation labeled $r_{Y1.2}$ and $r_{Y2.1}$.

Differences arise between the simple correlation coefficient between Y and a single independent variable, say X_1, and the partial correlation coefficient of Y and X_1. The difference is attributable to the correlation between the independent variable X_1 and any or all of the others. For complete multiple correlation analysis, all simple correlation coefficients for each pair of variables should be examined. Computer programs usually provide them in the form of a table or matrix with only one-half the matrix, either above or below the main diagonal, being needed.

Inferences about the Coefficients

The assumptions of multiple linear regression inference procedures are

1. For any given values of the independent variables, the actual dependent values are normally distributed about the regression plane.

2. The standard deviations (and variances) of these distributions around Y_c are equal.

3. The deviation of each dependent value of Y from the regression plane is independent of the deviation of the others.

Inference about Multiple Regression Coefficients

The population multiple regression equation is expressed as

$$\mu_{Y \cdot 12} = \alpha + \beta_1 X_1 + \beta_2 X_2.$$

Estimation of the Regression Coefficients

The standard errors of b_1 and b_2 are needed for interval estimators of β_1 and β_2. These errors are termed s_{b_1} and s_{b_2}. The confidence interval for each β_i can then be determined.

Hypothesis Testing about Regression Coefficients

The test statistic for testing a hypothesis about the regression coefficient β_i is

$$T_i = \frac{b_i - \beta_i}{s_{b_i}}$$

with critical t values used for decision-making purposes.

Inference about the Partial Correlation Coefficients

The test statistics for testing hypotheses about the partial correlation coefficients ρ_1 and ρ_2 are

$$T_1 = r_{Y1.2} \sqrt{\frac{n - k - 1}{1 - r_{Y1.2}^2}} \quad \text{and} \quad T_2 = r_{Y2.1} \sqrt{\frac{n - k - 1}{1 - r_{Y2.1}^2}}.$$

KEY TERMS

Canned computer program	A prewritten computer program used to relieve an investigator from the extensive calculations used in determining values of statistics needed for inferential decision making. An attempt is made by the programmer to make the data input simple and the output easily interpreted.
Coefficient of multiple determination	The multiple regression counterpart of the coefficient of determination in simple linear regression. It is a measure of the proportion of the total variation of the dependent variable that is explained or accounted for by the regression equation.
Collinearity	A complication in multiple regression analysis caused by intercorrelation between the independent variables. Also can be multicollinearity.
Hyperplane	The geometric interpretation of the regression equation with three or more independent variables. One variable produces a straight line: two variables can be graphed as a plane: but three or more variables must be abstractly thought of as a hyperplane that can be expressed mathematically but cannot be graphed.
Normal equations	The set of equations in the regression coefficients arrived at by applying the least squares procedure. The solution of the set of normal equations obtains the values for the regression equation coefficients.
Partial coefficient of correlation	A measure of the strength of relationship between the dependent variable and any one of the independent variables while the others are held constant. This measure may differ from a simple correlation coefficient because of collinearity.
Partial regression coefficients	The multipliers of the independent variables in the regression equation. They represent the net change in the dependent variable for a unit change in their respective independent variable.

Standard error of   In multiple regression analysis, the
the estimate   measure of the variation of Y from points
                      on the calculated regression plane or
                      hyperplane.

FORMULA REVIEW

(16-1) The linear multiple regression equation

$$Y_c = a + b_1 X_1 + b_2 X_2 + \cdots + b_k X_k$$

In the following formulas this shortcut in notation is used.

$$\Sigma x^2 = \sum_{i=1}^{n} (X_i - \bar{X})^2 \quad \text{(the variation of X)}$$

$$\Sigma y^2 = \sum_{i=1}^{n} (Y_i - \bar{Y})^2 \quad \text{(the variation of Y)}$$

$$\Sigma xy = \sum_{i=1}^{n} (X_i - \bar{X})(Y_i - \bar{Y}) \quad \text{(the covariation of X, Y)}$$

Regression coefficients

$$b_1 = \frac{(\Sigma x_2^2 \ \Sigma x_1 y) - (\Sigma x_2 y \ \Sigma x_1 x_2)}{D}$$

$$b_2 = \frac{(\Sigma x_1^2 \ \Sigma x_2 y) - (\Sigma x_1 y \ \Sigma x_1 x_2)}{D}$$

where

$$D = \Sigma x_1^2 \ \Sigma x_2^2 - (\Sigma x_1 x_2)^2$$

$$a = \bar{Y} - b_1 \bar{X}_1 - b_2 \bar{X}_2$$

Standard error of estimate for two independent variables

$$s_{Y \cdot 12} = \sqrt{\frac{\Sigma y^2 - (b_1 \ \Sigma x_1 y + b_2 \ \Sigma x_2 y)}{n - 3}}$$

Coefficient of multiple determination

$$R_{Y \cdot 12}^2 = \frac{b_1 \ \Sigma x_1 y + b_2 \ \Sigma x_2 y + \cdots + b_k \Sigma x_k y}{\Sigma y^2}$$

Partial coefficient of correlation

$$r_{Y1 \cdot 2}^2 = \sqrt{1 - \frac{1 - R_{Y \cdot 12}^2}{1 - r_{Y2}^2}} = \frac{r_{Y1} - r_{Y2} r_{12}}{\sqrt{1 - r_{Y2}^2} \ \sqrt{1 - r_{12}^2}}$$

$$r_{Y2.1}^2 = \sqrt{1 - \frac{1 - R_{Y \cdot 12}^2}{1 - r_{Y1}^2}} = \frac{r_{Y2} - r_{Y1}r_{12}}{\sqrt{1 - r_{Y1}^2}\sqrt{1 - r_{12}^2}}$$

where r_{Y1}, r_{Y2}, and r_{12} are the simple correlation coefficients of the respective pairs of variables.

Standard errors of b_1 and b_2 $\quad s_{b_1} = \dfrac{s_{Y \cdot 12}}{\sqrt{(\Sigma x_1^2)(1 - r_{12}^2)}} \quad s_{b_2} = \dfrac{s_{Y \cdot 12}}{\sqrt{(\Sigma x_2^2)(1 - r_{12}^2)}}$

Test statistic for testing hypothesis for the population coefficients β_i $\quad T_i = \dfrac{b_i - \beta_i}{s_{b_i}}$

Test statistics for testing hypotheses about the partial correlation coefficients ρ_1 and ρ_2 $\quad T_1 = r_{Y1.2}\sqrt{\dfrac{n - k - 1}{1 - r_{Y1.2}^2}} \quad T_2 = r_{Y2.1}\sqrt{\dfrac{n - k - 1}{1 - r_{Y2.1}^2}}$

ORGANIZED LEARNING QUIZ

1. The use of a prediction or a confidence interval is based on the assumption that the conditional distributions of Y are

    a. the same as the conditional distributions of the X_i's.
    b. never negative.
    c. normal and have equal means.
    d. normal and have equal variances.

2. The primary objective of multiple regression analysis is to

    a. predict the value of multiple variables, given the value of an associated variable.
    b. predict the value of one variable, given the value of an associated variable.
    c. predict the value of one variable, given the value of multiple associated variables.
    d. predict the value of multiple variables, given the value of multiple associated variables.

3. The multiple regression equation is determined by the process of

   a. least squares.
   b. the central limit theorem.
   c. correlation analysis.
   d. minimizing sums of errors.

4. In linear multiple regression analysis the constant a refers to

   a. the X intercept.
   b. the value of Y when all X_i's are equal to zero.
   c. the maximum value of the dependent variable Y.
   d. the change in Y for every unit change of the X_i's.

5. Each of the b_i multiple regression coefficients is

   a. the conditional slope between the ith independent variable and the dependent variable, given that the other independent variables are held constant.
   b. never negative.
   c. proportional to the importance of its respective ith independent variable in predicting a value for Y.
   d. normally distributed with mean equal to 1/k.

6. In multiple regression analysis the standard error of the estimate

   a. is determined by considering pairs of variables, one pair at a time.
   b. is determined by considering all independent variables as the basis of the conditional standard deviation.
   c. is the square root of the sums of the squares of the deviations from the regression equation.
   d. is calculated exactly the same as in simple linear regression analysis.

7. The coefficient of multiple determination is

   a. indicative of the extent of the relationship of the dependent variable to the independent variable.
   b. labeled R instead of r, as for the simple coefficient of correlation.
   c. indicative of the proportion of variation in the dependent variable explained by the knowledge of the independent variables.
   d. displayed in matrix form by computer programs.

8. The statistic that indicates the correlation between one of the independent variables in the multiple correlation analysis and the dependent variable, with the other independent variable(s) held constant, is

   a. the multiple regression coefficient.
   b. the coefficient of determination.
   c. the coefficient of correlation.
   d. the coefficient of partial correlation.

9. When two independent variables are highly correlated, a complication in analysis occurs, named

   a. autocorrelation.
   b. homeoscedasticity.
   c. collinearity.
   d. stepwise correlation.

10. A correlation matrix gives the values of

    a. all partial correlation coefficients between pairs of variables.
    b. all simple correlation coefficients between pairs of variables.
    c. the coefficient of determination.
    d. the coefficient of multiple correlation.

11. In computing the standard error of the estimate when there are two independent variables, the number of degrees of freedom is _____.

12. The sums of the cross products of two variables is called the _____ of the two variables.

13. Simple correlation coefficients sometimes are quite different from partial correlation coefficients due to _____. (one word)

14. To make inferences about the regression coefficients of the independent variable, it is necessary first to compute the values of the standard errors of _____ and _____.

15. In multiple linear regression analysis the least squares criterion requires that _____ be a minimum.

16. What is meant by the term linearity?

    _____
    _____
    _____

17. Why is it possible to have X_1 be a measure of different units than X_2 when they are summed to equal units of Y?

    _____
    _____
    _____

18. A regression coefficient that is positive in sign in a simple linear regression equation may change to a negative sign for the same independent variable in a multiple regression equation. Why?

    _____

19. Which computations make a computer desirable when including a large number of independent variables in the analysis?

20. Using the term variation, describe the computational form of the standard error of the estimate in two variable multiple regression analysis.

ANSWERS TO ORGANIZED LEARNING QUIZ

1. d   2. c
3. a   4. b
5. a   6. b
7. c   8. d
9. c   10. b

11. $n - 3$

12. covariation

13. collinearity

14. b_1 and b_2

15. $(Y - Y_c)^2$ the unexplained variation

16. Linearity is a description of a relationship between two variables. When the movement of one variable is associated with a constant proportional movement of the other, such that the relationship graphs as a straight line, the relationship is linear. Mathematically there exists an a such that $Y + a/X$ is constant for all occurring pairs of X and Y.

17. The units of the partial regression coefficients reduce each independent variable to a measure of units in Y. For example, Y can be food expenditures in dollars, X_1 an amount of income in dollars, and X_2 the size of a family. The coefficient of $X_1(b_1)$ has as units food expenditure in dollars per income in dollars, while b_2 is food expenditure in dollars per person size of a family.

18. One or more of the other independent variables in the multiple regression situation may be highly correlated with the given variable. The collinearity that occurs makes the partial regression coefficients highly unreliable for predicting influences.

19. The normal equations become increasingly difficult to compute, as do all other measures mentioned in this chapter. The computer also allows tests to be made that determine the most effective choice of variables to use when many are available.

20. The standard error of the estimate is the square root of the total variation of the dependent variable, minus the sum of the variation explained by the first independent variable and the variation explained by the second, divided by n minus three.

SAMPLE EXERCISE

Housing starts in a geographical area are thought to have a functional relationship with interest rates and number of marriages. Data on 12 areas taken at random at different period of times are presented below. Give a complete multiple linear regression analysis of the data.

Sample Number	Mortgage Interest Rates	Number of Marriages in Previous 12 Months	Housing Starts In Previous Month
1	6.25	224	755
2	7.75	221	580
3	6.00	232	790
4	6.50	231	665
5	8.00	214	540
6	8.50	203	515
7	6.00	227	690
8	5.75	245	810
9	5.50	235	835
10	6.50	234	645
11	6.75	238	610
12	7.50	232	595

Computational Table

X_1	X_2	Y	$X_1 * X_2$	$X_1 * Y$	$X_2 * Y$	X_1 SQ	X_2 SQ	Y SQ
6.25	224	755	1400	4718.75	169120	39.0625	50176	570025
7.75	221	580	1712.75	4495	128180	60.0625	48841	336400
6.00	232	790	1392	4740	183280	36.	53824	624100
6.50	231	665	1501.5	4322.5	153615	42.25	53361	442225
8.00	214	540	1712	4320	115560	64.	45796	291600
8.50	203	515	1725.5	4377.5	104545	72.25	41209	265225
6.00	227	690	1362	4140	156630	36.	51529	476100
5.75	245	810	1408.75	4657.5	198450	33.0625	60025	656100
5.50	235	835	1292.5	4592.5	196225	30.25	55225	697225
6.50	234	645	1521	4192.5	150930	42.25	54756	416025
6.75	238	610	1606.5	4117.5	145180	45.5625	56644	372100
7.50	232	595	1740	4462.5	138040	56.25	53824	354025
81	2736	8030	18374.5	53136.25	1839755	557	625210	5501150
6.75	228	669	18468	54202.5	1830840	546.75	623808	5373408
			−93.5	−1066.25	8915	10.25	1402	127742
			Ⓐ	Ⓑ	Ⓒ	Ⓓ	Ⓔ	Ⓕ

Explanation of Computational Table Results

Ⓐ   Covariation of X1, X2: $\Sigma\ x_1 x_2$

Ⓑ   Covariation of X1, Y: $\Sigma\ x_1 y$

Ⓒ   Covariation of X2, Y: $\Sigma\ x_2 y$

Ⓓ   Variation of X1: $\Sigma\ x_1^2$

Ⓔ   Variation of X2: $\Sigma\ x_2^2$

Ⓕ   Variation of Y: $\Sigma\ y^2$
(total variation)

Regression Coefficients

$$b_1 = \frac{E * B - C * A}{D * E - A * A} = \underline{-117.5}$$

$$b_2 = \frac{D * C - B * A}{D * E - A * A} = \underline{-1.478}$$

Standard error of estimate

$$s_{Y \cdot 12} = \frac{F - (b_1 * B + b_2 * C)}{n - 3} = \underline{41.67}$$

Coefficient of multiple determination

$$R^2_{Y.12} = \frac{b_1 * B + b_2 * C}{F} = \underline{.8777}$$

Y intercept

$$\bar{Y} - b_1 \bar{X}_1 - b_2 \bar{X}_2 = \underline{1799.17}$$

CASE PROBLEM

Richenburg Motors

Mr. John Richenburg is the owner of a foreign automobile agency in a town in Oregon of 90,000 population. In business for 18 years, he has experienced several stages of growth and development. The first years were rough in that growth was largely by word of mouth that his service was sincere, complete, and reliable.

In the fourth year of operations his sales of new cars began a sizable growth pattern that lasted for another 6 or 7 years. He attributed this growth to his continued sincere desire to treat the customer fairly; however, he realized that his firm's name had become a household word in the community because of his brother's record on the sports car circuits. Richenburg Motors had sponsored his brother in many local, regional, and national meets and he was a national point leader in his class for the make of automobiles they sold.

During the last few years John felt that the business had entered a more stable, mature phase of its development pattern and was grateful to the community for its support during past years. His method of advertising had changed recently, both in media used and in content. His primary means of advertising has shifted from classified newspaper displays to 1-minute spots on the two local television stations. His television spots were of two types, which he classified as being product oriented and educational oriented. The latter was of a public-service variety and ranged from mechanical maintenance tips to good driving reminders during particular times, such as ice and snow season, opening of school, fog, and so on.

John wonders whether there is a relationship between his sales and his advertising efforts. To explore this subject further, he made an appointment to see the father of one of the ladies in his credit department. The father is employed as an actuary of a life insurance company based in Oregon.

The suggestion was made to John that he make use of a multiple regression model; and he was given the necessary theoretical background and formulas to do the analysis. John has gathered the data in Exhibit A and intends to use the data in his calculations.

At the interview the actuary cautioned John on many points. These points John had transcribed from a recording of his meeting.

"A great amount of care must be used when developing a multiple regression model. Knowledge of the field, logical judgment, and theoretical analysis must be used when selecting the final variables that appear in the final predicting equation. Also, it is possible that some relevant variables may not be easily quantifiable.

"The restrictions to a linear model I have given you may not be serious if the range of variables is small; however, a plot of each of the independent variables with the dependent variable may be wise.

"Even though more variables added to the regression model may result in a better fit, consideration must be given to the cost tradeoff of the difficulties in acquiring more data and the increase in calculations necessary.

"A final caution given is one of the dangers of extrapolation. Statistically valid predictions are shaky at the best when using predictors outside the range of values used in the determination of the model."

Exhibit A

Sample Number	Sales	Minute TV Spots of Education Orientation	Minute TV Spots of Product Orientation
1	66	128	260
2	64	123	238
3	62	119	252
4	60	110	240
5	58	106	230
6	56	114	222
7	54	100	214
8	52	102	208
9	53	98	200
10	55	103	210
11	57	113	240
12	59	118	230
13	61	123	220
14	63	128	252
15	65	125	264

COMPUTER PROGRAM

The following BASIC program is for two-variable regression. Sample runs are given for the Sample Exercise and the Case Problem.

```
10 REM************************************************************
20 REM
30 REM      TITLE:       MULTREG
40 REM      LANGAUGE:    BASIC
50 REM      DESCRIPTIONS:
60 REM         THIS PROGRAM IS FOR TWO INDEPENDENT VARIABLES
70 REM           REGRESSION.
80 REM      INSTRUCTIONS:
90 REM         ENTER DATA AS REQUESTED BY THE PROGRAM.
100 REM     PROGRAMMER:
110 REM        GROVER WM. RODICH, PORTLAND STATE UNIV.
120 REM
130 REM************************************************************
140 KEY OFF
150  DIM X1(50),X2(50),Y(50)
```

```
160  CLS
170  PRINT "MULTIPLE REGRESSION PROGRAM FOR TWO INDEPENDENT VARIABLES ONLY"
180  PRINT:PRINT"         INPUT 0 TO EXIT PROGRAM"
190  PRINT "        INPUT 1 TO CONTINUE":PRINT:PRINT "YOUR CHOICE";
200  INPUT G
210  IF G=0 THEN 1500
220  IF G<>1 THEN 180
230  CLS
240  PRINT "WHAT IS THE SAMPLE SIZE?"
250  PRINT "N= ";
260  INPUT N
270  PRINT "INPUT YOUR VARIABLES THREE AT A TIME SEPARATED BY COMMAS"
280  PRINT "KEEP IN THE ORDER X(1), X(2), Y "
290  FOR I=1 TO N
300  INPUT X1(I),X2(I),Y(I)
310  NEXT I
320  PRINT
330  CLS:PRINT
340  PRINT TAB(17);"X1";TAB(31);"X2";TAB(45);"Y"
350  PRINT "     ","--------------------------------------------"
360  PRINT
370  T1=0
380  T2=0
390  T3=0
400  FOR I=1 TO N
410  T1=T1+X1(I)
420  T2=T2+X2(I)
430  T3=T3+Y(I)
440  PRINT "     ",X1(I),X2(I),Y(I)
450  NEXT I
460  PRINT "     ","--------------------------------------------"
470  PRINT
480  PRINT"TOTALS",T1,T2,T3
490  PRINT
500  M1=T1/N
510  M2=T2/N
520  M3=T3/N
530  PRINT "MEANS",M1,M2,M3
540  PRINT
550  PRINT
560  LOCATE 24,1:PRINT "TO CONTINUE OUTPUT, PRESS ANY KEY";:A$=INPUT$(1)
570  PRINT
580  CLS
590  PRINT
600  PRINT " "; "X1*X2";TAB(12);"X1*Y";TAB(24);"X2*Y";
610  PRINT TAB(36);"X1 SQ";TAB(48);"X2 SQ";TAB(60);"Y SQ"
620  PRINT "-----------------------------------------------------------"
630  PRINT
640  T4=0
650  T5=0
660  T6=0
670  T7=0
680  T8=0
690  T9=0
700  FOR I=1 TO N
710  T4=T4+X1(I)*X2(I)
720  T5=T5+X1(I)*Y(I)
730  T6=T6+X2(I)*Y(I)
740  T7=T7+X1(I)^2
750  T8=T8+X2(I)^2
760  T9=T9+Y(I)^2
770  PRINT X1(I)*X2(I);TAB(12);X1(I)*Y(I);TAB(24);X2(I)*Y(I);
780  PRINT TAB(36);X1(I)^2;TAB(48);X2(I)^2;TAB(60);Y(I)^2
790  NEXT I
800  PRINT "-----------------------------------------------------------"
810  PRINT T4;TAB(12);T5;TAB(24);T6;TAB(36);
820  PRINT T7;TAB(48);T8;TAB(60);T9
830  PRINT
840  PRINT M1*T2;TAB(12);M1*T3;TAB(24);M2*T3;TAB(36);
850  PRINT M1*T1;TAB(48);M2*T2;TAB(60);M3*T3
860  PRINT "-----------------------------------------------------------"
870  C1=T4-M1*T2
880  C2=T5-M1*T3
890  C3=T6-M2*T3
900  V1=T7-M1*T1
910  V2=T8-M2*T2
920  V3=T9-M3*T3
930  PRINT C1;TAB(12);C2;TAB(24);C3;TAB(36);
940  PRINT V1;TAB(48);V2;TAB(60);V3
950  PRINT
960  D=V1*V2-C1^2
970  B1=(V2*C2-C3*C1)/D
```

```
980   B2=(V1*C3-C2*C1)/D
990   A=M3-B1*M1-B2*M2
1000  LOCATE 24,1:PRINT "TO CONTINUE OUTPUT, PRESS ANY KEY":A$=INPUT$(1)
1010  PRINT
1020  CLS
1030  PRINT
1040    PRINT"THE REGRESSION EQUATION IS: Y=";A;"+";B1;"X1+";B2;"X2"
1050    PRINT
1060    PRINT "S(Y,12) =";((V3-B1*C2-B2*C3)/(N-3))^.5
1070    PRINT
1080    PRINT"R SQ(Y,12) =";(B1*C2+B2*C3)/V3
1090    PRINT
1100    PRINT"TABLE OF SIMPLE CORRELATION COEFFICIENTS"
1110  PRINT TAB(15);"X1";TAB(30);"X2";TAB(45);"Y"
1120    PRINT TAB(10);"----------------------------------------------"
1130    R1=C1/((V1*V2)^.5)
1140    R2=C2/((V1*V3)^.5)
1150    R3=C3/((V2*V3)^.5)
1160    PRINT "X1";TAB(10);"1.000"
1170    PRINT
1180    PRINT"X2";TAB(10);R1;TAB(25);"1.000"
1190    PRINT
1200    PRINT "Y";TAB(10);R2;TAB(25);R3;TAB(40);"1.000"
1210    PRINT
1220  LOCATE 24,1:PRINT"TO CONTINUE OUTPUT, PRESS ANY KEY":A$=INPUT$(1)
1230    PRINT
1240  CLS
1250    PRINT
1260    R4=(R2-R3*R1)/(((1-R3^2)^.5)*((1-R1^2)^.5))
1270    R5=(R3-R2*R1)/(((1-R2^2)^.5)*((1-R1^2)^.5))
1280    S1=(((V3-B1*C2-B2*C3)/(N-3))/(V1*(1-R1^2)))^.5
1290    S2=(((V3-B1*C2-B2*C3)/(N-3))/(V2*(1-R1^2)))^.5
1300    PRINT "R(Y1.2) = ";R4
1310    PRINT
1320    PRINT "R(Y2.1) = "; R5
1330    PRINT
1340    PRINT
1350    PRINT "S(B1) = "; S1
1360    PRINT
1370    PRINT "S(B2) = "; S2
1380    PRINT
1390    PRINT
1400    PRINT"THE TEST STATISTICS FOR THE REGRESSION COEFFIENTS ARE"
1410    PRINT "T(1) =";B1/S1,"T(2) ="; B2/S2
1420    PRINT
1430    PRINT"THE TEST  STATISTICS FOR THE PARTIAL CORRELATION COEF-"
1440    PRINT "FICIENTS ARE"
1450    PRINT "T(1)=";R4*((N-3)/(1-R4^2))^.5;"      ";
1460    PRINT "T(2)= ";R5*((N-3)/(1-R5^2))^.5
1470  PRINT:PRINT
1480  LOCATE 24,1:PRINT "TO CONTINUE PRESS ANY KEY" :A$=INPUT$(1)
1490  GOTO 160
1500  KEY ON:END




MULTIPLE REGRESSION PROGRAM FOR TWO INDEPENDENT VARIABLES ONLY

       INPUT 0 TO EXIT PROGRAM
       INPUT 1 TO CONTINUE

YOUR CHOICE? 1


WHAT IS THE SAMPLE SIZE?
N= ? 12
INPUT YOUR VARIABLES THREE AT A TIME SEPARATED BY COMMAS
KEEP IN THE ORDER X(1), X(2), Y
? 6.25,224,755
? 7.75,221,580
? 6,232,790
? 6.5,231,665
? 8,214,540
? 8.5,203,515
? 6,227,690
? 5.75,245,810
? 5.5,235,835
? 6.5,234,645
? 6.75,238,610
? 7.5,232,595
```

	X1	X2	Y
	6.25	224	755
	7.75	221	580
	6	232	790
	6.5	231	665
	8	214	540
	8.5	203	515
	6	227	690
	5.75	245	810
	5.5	235	835
	6.5	234	645
	6.75	238	610
	7.5	232	595
TOTALS	81	2736	8030
MEANS	6.75	228	669.1667

TO CONTINUE OUTPUT, PRESS ANY KEY

X1*X2	X1*Y	X2*Y	X1 SQ	X2 SQ	Y SQ
1400	4718.75	169120	39.0625	50176	570025
1712.75	4495	128180	60.0625	48841	336400
1392	4740	183280	36	53824	624100
1501.5	4322.5	153615	42.25	53361	442225
1712	4320	115560	64	45796	291600
1725.5	4377.5	104545	72.25	41209	265225
1362	4140	156630	36	51529	476100
1408.75	4657.5	198450	33.0625	60025	656100
1292.5	4592.5	196225	30.25	55225	697225
1521	4192.5	150930	42.25	54756	416025
1606.5	4117.5	145180	45.5625	56644	372100
1740	4462.5	138040	56.25	53824	354025
18374.5	53136.25	1839755	557	625210	5501150
18468	54202.5	1830840	546.75	623808	5373409
-93.5	-1066.25	8915	10.25	1402	127741.5

TO CONTINUE OUTPUT, PRESS ANY KEY

THE REGRESSION EQUATION IS: $Y = 1799.17 + -117.5019\ X1 + -1.47748\ X2$

$S(Y,12) = 41.66913$

$R\ SQ(Y,12) = .8776681$

TABLE OF SIMPLE CORRELATION COEFFICIENTS

	X1	X2	Y
X1	1.000		
X2	-.7799663	1.000	
Y	-.9318182	.6661649	1.000

TO CONTINUE OUTPUT, PRESS ANY KEY

R(Y1.2) = -.8832147

R(Y2.1) = -.2669081

S(B1) =   20.79706

S(B2) =   1.778238

THE TEST STATISTICS FOR THE REGRESSION COEFFIENTS ARE
T(1) =-5.649927   T(2) =-.8308671

THE TEST STATISTICS FOR THE PARTIAL CORRELATION COEF-
FICIENTS ARE
T(1)=-5.649936   T(2)= -.8308665

TO CONTINUE PRESS ANY KEY

MULTIPLE REGRESSION PROGRAM FOR TWO INDEPENDENT VARIABLES ONLY

      INPUT 0 TO EXIT PROGRAM
      INPUT 1 TO CONTINUE

YOUR CHOICE? 1

WHAT IS THE SAMPLE SIZE?
N= ? 15
INPUT YOUR VARIABLES THREE AT A TIME SEPARATED BY COMMAS
KEEP IN THE ORDER X(1), X(2), Y
? 128,260,66
? 123,238,64
? 119,252,62
? 110,240,60
? 106,230,58
? 114,222,56
? 100,214,54
? 102,208,52
? 98,200,53
? 103,210,55
? 113,240,57
? 118,230,59
? 123,220,61
? 128,252,63
? 125,264,65

	X1	X2	Y
	128	260	66
	123	238	64
	119	252	62
	110	240	60
	106	230	58
	114	222	56
	100	214	54
	102	208	52
	98	200	53
	103	210	55
	113	240	57
	118	230	59
	123	220	61
	128	252	63
	125	264	65
TOTALS	1710	3480	885
MEANS	114	232	59

TO CONTINUE OUTPUT, PRESS ANY KEY

33280	8448	17160	16384	67600	4356
29274	7872	15232	15129	56644	4096
29988	7378	15624	14161	63504	3844
26400	6600	14400	12100	57600	3600
24380	6148	13340	11236	52900	3364
25308	6384	12432	12996	49284	3136
21400	5400	11556	10000	45796	2916
21216	5304	10816	10404	43264	2704
19600	5194	10600	9604	40000	2809
21630	5665	11550	10609	44100	3025
27120	6441	13680	12769	57600	3249
27140	6962	13570	13924	52900	3481
27060	7503	13420	15129	48400	3721
32256	8064	15876	16384	63504	3969
33000	8125	17160	15625	69696	4225
399052	101488	206416	196454	812792	52495
396720	100890	205320	194940	807360	52215
2332	598	1096	1514	5432	280

TO CONTINUE OUTPUT, PRESS ANY KEY

THE REGRESSION EQUATION IS: Y= 8.610476 + .248567 X1+ 9.505554E-02 X2

S(Y,12) = 1.504883

R SQ(Y,12) = .9029426

TABLE OF SIMPLE CORRELATION COEFFICIENTS

	X1	X2	Y
X1	1.000		
X2	.8131785	1.000	
Y	.9184581	.8886929	1.000

TO CONTINUE OUTPUT, PRESS ANY KEY

R(Y1.2) = .7337015

R(Y2.1) = .616091

S(B1) = 6.645165E-02

S(B2) = 3.508238E-02

THE TEST STATISTICS FOR THE REGRESSION COEFFIENTS ARE
T(1) = 3.74057   T(2) = 2.709496

THE TEST STATISTICS FOR THE PARTIAL CORRELATION COEF-
FICIENTS ARE
T(1)= 3.740576   T(2)= 2.709495

TO CONTINUE PRESS ANY KEY

MULTIPLE REGRESSION PROGRAM FOR TWO INDEPENDENT VARIABLES ONLY

    INPUT 0 TO EXIT PROGRAM
    INPUT 1 TO CONTINUE

YOUR CHOICE? 0

OK

17

TIME SERIES

CHAPTER LEARNING OBJECTIVES

On completion of this chapter the student should be able to

- Explain how statistical inference procedures can be useful in forecasting.

- List examples of time series in both business and nonbusiness situations.

- List the components of a time series studied when the series is decomposed.

- Explain the difference between a multiplicative and an additive general time-series model.

- Explain the essential differences between linear regression analysis and linear trend analysis.

- Explain how seasonal indexes are computed and used.

- Smooth a time-series data set by the method of exponential smoothing and explain the consequence of the choice of the smoothing constant.

- Explain the economic use of index numbers, listing their advantages over the raw data.

SYNOPSIS

Decision making, planning, and establishing goals and objectives for strategic purposes usually require a certain hint of what the future might bring in terms of the variables important in the environment of a firm. The application of statistical inference about future events, based on samples of present and past occurrences, is helpful in such situations. When the occurrences are graphed, with time being the horizontal axis, the result is called a time series. Gross revenue for a firm plotted monthly over several months is an example of a time series. So are weekly rainfall, accidents in a city during a 24-hour period, high-tide levels, and any other measurements plotted over a period of time.

The statistician is interested in any pattern that might be exhibited in the sample data so that extrapolations can be made toward the future for forecasting purposes. A strong assumption is that past behavior will reasonably continue in the future.

Components of Time Series

Secular Trend. A secular (long-term) trend is manifested by a steady and gradual change in movements of a variable. Examples include long-term increase in population and industrial output of the nation.

Cyclical variations. Two-year or longer repetitive movements, such as business cycles, changes in interest rates, and housing starts, are termed cyclical variations.

Seasonal Variations. Periodic variations that are repetitive over a period of one year or less are called seasonal variations. Soft drink sales, ski resort revenues and boat sales follow patterns that are seasonal in nature. Seasonal variations are much more predictable than cyclical variations.

Irregular Variations. After a time series has been decomposed and the three variations above accounted for, the residual is random and largely unexplained. This kind of variation is called irregular variation. It consists of erratic, short, unpredictable movements in the time-series data. It is similar to the waves of an ocean as compared to the more predictable, steady tidal movements.

Decomposition of Time Series

A time-series model attempts to specify the relationship each of the four components above has with the series as a whole. Multiplicative models and additive models have been accepted as having good approximations to the true component relationships. When using either a multiplicative or an additive model, the implicit assumption is the independence of each component. This gives a first approximation; and if the assumption is not realistic, then results can be modified at a later time.

Analysis of Trend

Linear Trend. Least squares can be used to fit a line to time-series data; however, the deviations of the actual data from the trend line contain results of the other components and are not just random errors as in regression analysis. The trend-line equation uses time as the independent variable and the trend value is the dependent variable.

Curvilinear Trend. Trends can follow a pattern that is curvilinear rather than linear. Examples are exponential changes, S-shaped growth curves, and second-degree curves.

By eliminating all components except trend through the method of moving averages, the trend pattern is approximated.

Cyclical Variations

If the data are annual figures, then there will be no seasonal variations and the time series will consist of only the other three. Also, since annual data are used, the irregular component has had a long enough time to cancel itself or be smoothed out. Therefore annual data are primarily affected by trend and cycles. If a multiplicative model is used, cyclical relatives can be computed by dividing the model by the trend factor.

Seasonal Variations

For operational planning of one year or less, the seasonal pattern is important. Seasonal indexes are computed for each season to measure their variation. When utilized, the resulting data are deseasonalized or seasonly adjusted.

Construction of Seasonal Index. The ratio-to-moving-average method obtains a series of moving averages that contains approximately only trend and cyclical components. This series is then divided into the original series to cancel out the trend and cyclical components, leaving only the seasonal and irregualr components. The irregular components are approximately removed by averaging quarters (if quarterly data are used) or months (if monthly data are used). What remains are unadjusted seasonal indexes. They are unadjusted because they do not necessarily total to 400 for quarterly data or 1200 for monthly data. They are adjusted by being multiplied by the ratio necessary to make them sum to 400 (or 1200).

Use of Seasonal Indexes. The original series becomes deseasonalized by dividing it by the seasonal index. The resulting seasonally adjusted data can now be investigated for trend and cycle components.

Forecasting

Trend Forecast. The linear trend line can be used directly to predict or extrapolate the given data into the future. Multiple regression is applicable if more than two independent variables are involved in the forecast. The forecaster must remember the critical assumption of this technique of basing the future on relationships that occurred in the past.

Seasonal Forecast. After the long-term trend component has been determined, a seasonal adjustment can be made for the shorter periods within a year.

Cyclical Forecast. Long-term economic conditions are more difficult to predict than trend or seasonal movements. The performance of certain variables, such as the Gross National Product, follow the conditions of the economy fairly closely. These variables are called economic indicators and forecasters watch them closely to try to detect significant movements that possibly give rise to changes in business conditions. Economic indicators can be leading, coincident, or lagging in nature when compared to the movements of general economic activity.

Exponential Smoothing

When no significant trend is apparent, the method of exponential smoothing is useful. The technique allows previous values of a variable to be considered in a weighted average where the weights are reduced in an exponential fashion as time goes back in history. The parameter α allows the investigator to determine the speed in which the exponential change in the weights occurs. Small values of α put more weight on past values; sudden changes in the recent past have little effect on the forecast. Large values of α (close to one) put most of the weight on recent data; therefore the model is more sensitive to rapid changes.

Index Numbers

Index numbers are given time measures expressed as a percentage of the measure at a base period. When using composite index numbers, a method is needed to determine how the items are to be averaged. For composite price indexes, if all prices are expressed as dollars per pound, then a simple average might do. However, if some prices are per dozen, per units, or by the pound, then difficulties in averaging arise. This problem is usually solved by using comparable usage amounts as weights for averaging purposes. The results are then called a weighted price index.

To correct movements in economic indicators for price changes caused by inflation, the indicators are divided by the price index and multiplied by 100. The indicator has thus been deflated and gives what is called its real value based on constant dollars.

KEY TERMS

Additive time-series model	A time-series model that assumes the original series is the sum of the values of secular trend, seasonal component, cyclical component, and the irregular component.
Constant dollars	Dollars that have equal purchasing power. Goods valued in constant dollars are expressed in their real value as related to a base period.
Curvilinear trend	Curved trend lines that fit the observed data. Examples include exponential, growth, and second- and higher-degree curves.
Cyclical variations	Recurrent upward and downward movements around a trend line over long-term periods.
Economic indicators	General elements of the economy whose time series are frequently used for forecasting purposes.

Exponential smoothing	A modeling technique designed to weight the actual value and the forecast value for time t in order to obtain a forecast for time period t + 1.
Index numbers	The value of a given period expressed as a percentage of the value of a base period for the same variable.
Linear trend	The least squares line that fits the observed data over a long period of time, usually covering two or more cycles.
Moving average	A changing value that is the average of the last n periods.
Multiplicative time-series model	A time-series model that assumes the original series is the product of the values of secular trend, seasonal component, cyclical component, and the irregualr component.
Price index	The average of the price relatives for many commodities, times 100.
Price relative	A ratio of a current price to a base-year price for a single commodity.
Seasonal variations	Periodic, regular movements around a trend line over periods within a year. The movements are more regular (and more predictable) than cyclical movements and the time period is much shorter.
Secular trend	A long-time, steady, and gradual change in the value of a time-series variable.
Statistical deflation	The effects of price changes on the value of goods and services removed by the use of index numbers.
Time series	A chronological sequence of measurements for some variable or composite of variables.
Time-series model	The specification of the relationship of the time series as a whole to its various components. Two accepted approximation models are the multiplicative model and the additive model.
Weighted price index	The ratio of the sum of current weighted prices to the sum of base-year weighted prices, times 100, where the weights are specified quantities of commodities.

FORMULA REVIEW

(17-1) Multiplicative time-series model $\quad Y = T * S * C * I$

	where	Y = original value of time series
		T = value of secular trend
		S = value of seasonal component
		C = value of cyclical component
		I = value of irregular component

(17-3) Trend line

$$T = a + bX$$

where
T = trend value
X = time period
a, b = the least squares parameters

(17-7) Deseasonalized series value

$$= \frac{Y}{\text{seasonal index}} * 100$$

(17-8) Exponential smoothing

$$F_{t+1} = F_t + \alpha(A_t - F_t)$$
or $\alpha A_t + (1 - \alpha)F_t$ $(0 \leq \alpha \leq 1)$

where
F_{t+1} = new forecast for time period $t + 1$
F_t = old forecast made for time period t
A_t = actual value in time period t
α = exponential smoothing constant

(17-10) Price index

$$I = \frac{\Sigma (P_g/P_b)}{n} * 100$$

where
P_g = given period price
P_b = base period price
n = number of commodities under consideration

(17-11) Weighted price index

$$I = \frac{\Sigma P_q Q_b}{\Sigma P_q Q_b} * 100$$

where Q_b is the base period quantities used as weights

(17-12) Real value

$$= \frac{\text{current value}}{\text{appropriate price index}} * 100$$

ORGANIZED LEARNING QUIZ

1. When utilizing statistical inference procedures in forecasting, observed time-series data assume the role of the

    a. sample.
    b. population.
    c. statistic.
    d. sample size and degrees of freedom.

2. Gross National Product, national income, unemployment, Consumer Price Index, and a firm's revenue figures are all examples of

    a. index numbers.
    b. time series.
    c. lagging indicators.
    d. weighted price variables.

3. The components of a time series include

    a. index numbers and cyclical variations.
    b. secular trends and irregular components.
    c. seasonal components and regression lines.
    d. all the above.

4. The time horizon for secular trends normally spans

    a. less than 6 months.
    b. between 6 months and a year.
    c. between 1 and 2 years.
    d. longer than 2 years.

5. Residual variation and random variation are other names for

    a. irregular variation.
    b. regular variation.
    c. predictable variation.
    d. recurrent variation.

6. The time series of beer sales in Colorado would have a strong component of

    a. index numbers.
    b. forecasting smoothing constants.
    c. a seasonal value.
    d. none of the above.

7. Deviations of observed annual Y values from a time-series trend include random errors, as in regression analysis, but they also include

    a. cyclical components only.
    b. cyclical and seasonal components.
    c. seasonal components only.
    d. an attribute of normality with equal variances.

8. The 12 monthly seasonal indexes must total to

   a. 400.
   b. 100.
   c. 1200.
   d. 12.

9. Economic leading indicators are useful in determining the

   a. secular trend.
   b. the cyclical turning point.
   c. the seasonal variations.
   d. irregular component.

10. Even though the exponential smoothing equation is computed by using the forecasted value and the actual value for the current period to forecast the next period, the forecast is

    a. actually based on all past periods.
    b. actually based on all past periods of the current season.
    c. actually based on the error made in forecasting the next period.
    d. actually based on the sum of all past errors.

11. In exponential smoothing the only variable that needs to be stored from one period to the next is the _____.

12. The ratio of the given-year price to the base-year price is termed the _____.

13. By using _____, a time series can be deseasonalized. Deseasonalized data are called _____.

14. A time series is usually broken down into four components _____, _____, _____, and _____.

15. The length of a business cycle is measured from _____ to _____.

16. Explain how moving averages are used to obtain a model containing only seasonal and irregular components when both are largely absent in moving averages.

    _____

    _____

    _____

17. Explain the characteristics of the S-shaped curvilinear trend curve.

    _____

    _____

    _____

18. How is the moving average of quarterly data centered on a particular quarter instead of between the second and third quarter?

    _____

    _____

    _____

19. After a time series has been decomposed, how is a forecast made?

    _____

    _____

    _____

20. Why do you think exponential smoothing is not as useful with data containing a trend as with relatively stable series?

    _____

    _____

    _____

ANSWERS TO ORGANIZED LEARNING QUIZ

1. a   2. b
3. b   4. d
5. a   6. c
7. a   8. c
9. b   10. a

11. new forecast

12. price relative

13. seasonal indexes, seasonally adjusted

14. secular trend, cyclical fluctuations, seasonal variations, and irregular changes.

15. recession to recession or prosperity (boom) to prosperity

16. As stated, the moving average model primarily consists of trend and cyclical components. When the multiplicative model for the whole series is divided by the moving averages, the trend and cyclical components cancel and the seasonal and irregular components are left.

17. The S-shaped curve describes the growth pattern of a firm. The initial period of a firm is characterized by slow growth, followed by an increasing rate of growth, and finally, a slowing down of growth during the mature years.

18. The four-quarter moving average is centered on the third quar-

ter by averaging two consecutive moving averages.

19. The decomposition has determined the values for the trend, cycle, and seasonal components. The time-series approximation is then reconstructed by applying the cycle adjustments and the seasonal indexes to the trend line.

20. The forecast is based on all actual values of the past. If a trend is present, the forecast will lag seriously behind the trend. If wild flucuations and changes of direction are present in the series, the smoothing characteristic of the model will cause it to be slow to describe them or it will not describe them at all, since they will tend to average themselves out.

SAMPLE EXERCISE

The following data are usage figures for a particular expensive inventory part. Use exponential smoothing to generate a forecast for the 17th time period. Try values of .1, .3, .5, and .7 and select the value of α that generates the smallest mean square error.

(The answer to this exercise is in the computer-run listing at the end of this chapter.)

Time Period	Part Usage
1	19
2	15
3	39
4	102
5	90
6	29
7	90
8	46
9	30
10	66
11	80
12	89
13	82
14	17
15	26
16	29

CASE PROBLEM

The Scappoose Hospital

The Scappoose Hospital was constructed with federal WPA funds in 1935. Originally its capacity was 200 beds; however, in 1956 it expanded to 350 beds. At present there are over 500 personnel on the payroll, including doctors, nursing staff, dietary staff, laboratory technicians, custodial crew, and office personnel.

Hospital record keeping was computerized in 1964; and in 1971 the hospital purchased its present in-house small computer system. With the amount of unused computer time available, Tom Blodgett, the manager of computer services, has been able to suggest many ways to improve the information content needed for effective decision making by the hospital administrators.

One application Tom has developed is an on-line system for the front desk that works in a manner similar to an airline reservation system. By inquiry, the front desk attendant can determine in which room a patient is located, how long the stay has been, the doctor's name, and any out-of-the-ordinary conditions that may be present. The system responds to either an entry of a room number or an entry of a patient's name.

Mr. Blodgett has been asked by the office staff if he could be of any help in predicting the number of patients requesting bed space each day. He has made some preliminary investigations by inspecting the daily-entry rosters and by interviewing the persons who are in charge of the roster generation.

Preliminary Findings

Mr. Blodgett decided that the data he was able to obtain were basically a time series and that the problem was one of forecasting. Unsure of the particular model he should use, he felt that he would know more about the nature of the data after he made a time plot with time (in days) along the horizontal axis and the number of daily registrations along the vertical axis. The data he used are listed in Exhibit A.

Exhibit A
Daily Patient Registrations
Scappoose Hospital

Day No.	Registrations
1	48
2	59
3	58
4	62
5	50
6	49
7	57
8	81
9	78
10	73
11	58
12	43
13	50
14	44
15	27
16	38
17	37
18	74
19	64
20	59

After plotting the preceding data, Mr. Blodgett observed that each data point seemed to be highly correlated with the point immediately before it (lag one). If such was the case, perhaps the best forecast for tomorrow was simply today's actual figure. In an

exponential smoothing model this would correspond to a smoothing
constant of one where all the weight is put on the present actual
and no concern is made for past data.

To test his theory, he wrote a small exponential smoothing
program and tried ∝ values of 1.0, .9, and .8 to compare their
results. The output of his efforts are at the end of this chapter.
He now wondered what his next step should be. He had some time
before the office staff expected his recommendation; and other
forecasting models hadn't yet been tried.

COMPUTER PROGRAM

The following program written in BASIC allows the user to
simulate and search for an exponential smoothing constant for a
time-series data set. For a data set larger than 50, statement
240 needs to be modified.

```
10 REM*********************************************************
20 REM
30 REM      TITLE:        TIMESMOO
40 REM
50 REM      LANGAUGE:  BASIC
60 REM
70 REM      DESCRIPTIONS:
75 CLS
80 PRINT "   THE FOLLOWING PROGRAM ALLOWS THE USER"
90 PRINT "   TO SIMULATE AND SEARCH FOR AN EXPONENTIAL"
100 PRINT"   SMOOTHING CONSTANT FOR A TIME-SERIES"
110 PRINT"   DATA SET.  FOR A DATA SET LARGER THAN"
120 PRINT"   50, STATEMENT 240 NEEDS TO BE MODIFIED."
130 PRINT
135 PRINT:PRINT:PRINT
140 REM
150 REM      INSTRUCTIONS:
160 REM         ENTER DATA AS REQUESTED BY THE PROGRAM.
170 REM
180 REM      PROGRAMMER:
190 REM         GROVER WM. RODICH, PORTLAND STATE UNIV.
200 REM
210 REM*********************************************************
220 REM
230 KEY OFF
240 DIM A(50),F(51)
250 CLS
260 LOCATE 8,1
270 PRINT "THIS IS A SINGLE EXPONENTIAL SMOOTHING PROGRAM THAT WILL"
280 PRINT "ALLOW THE USER TO SIMULATE WITH DIFFERENT VALUES FOR THE"
290 PRINT "SMOOTHING CNSTANT (ALPHA)."
300 PRINT
310 PRINT "AFTER EACH CHOICE OF ALPHA THE STANDARD ERROR OF THE "
320 PRINT "RESIDUALS WILL BE GIVEN."
330 PRINT
340 PRINT "IT IS ASSUMED THAT THERE IS VERY LITLE TREND PRESENT "
350 PRINT "IN THE DATA.  THE INITIAL FORECAST EQUAL TO THE FIRST DATA"
360 PRINT "POINT WILL BE GENERATED BY THE PROGRAM."
370 PRINT
380 PRINT "HOW MANY DATA POINTS DO YOU HAVE?   N= ";
390 INPUT N
400 PRINT "ENTER THE ACTUAL VALUES ONE AT A TIME FOLLOWED BY A RETURN."
410 M=0
420 FOR I=1 TO N
430 INPUT A(I)
440 M=M+A(I)
450 NEXT I
460 M=M/N
470 PRINT
480 CLS:LOCATE 8,1
490 PRINT "THE MEAN OF THE DATA POINTS IS";M
500 PRINT "WHAT ALPHA DO YOU WISH TO TRY?   ALPHA =   ";
510 INPUT A1
520 PRINT
530 F(1)=A(1)
540 FOR I=1 TO N
550 F(I+1)=F(I)+A1*(A(I)-F(I))
```

```
560 NEXT I
570 R3=0
580 R4=0
590 FOR I=1 TO N
600 R3=R3+(A(I)-F(I))
610 R4=R4+(A(I)-F(I))^2
620 NEXT I
630 C=R3*R3/N
640 V=R4-C
650 V2=V/(N-1)
660 S=V2^.5
670 PRINT "THE STANDARD ERROR OF THE DEVIATIONS IS ";S
680 PRINT
690 PRINT "THE MSE IS ";R4/N
700 PRINT
710 PRINT
720 PRINT "DO YOU WISH TO TRY A DIFFERENT ALPHA?   TYPE 1 FOR YES "
730 PRINT "AND 0 (ZERO) FOR NO.        RESPONSE";
740 INPUT Q
750 IF Q=1 THEN 480
760 IF Q<>0 THEN 720
770 PRINT
780 PRINT
790 FOR I=1 TO N
800 PRINT "FORECAST(";I;") =";F(I)
810 NEXT I
820 PRINT
830 PRINT
840 PRINT "F(";N+1;" ) =";F(N+1)
850 PRINT
860 PRINT
870 PRINT "DO YOU WISH TO TRY ANOTHER ALPHA?"
880 PRINT "AGAIN, TYPE 1 FOR YES AND 0 FOR NO.       RESPONSE";
890 INPUT Q
900 IF Q=1 THEN 480
910 IF Q<>0 THEN 870
920 PRINT
930 PRINT "PROGRAM COMPLETED--THANK YOU"
940 KEY ON:END
```

THIS IS A SINGLE EXPONENTIAL SMOOTHING PROGRAM THAT WILL
ALLOW THE USER TO SIMULATE WITH DIFFERENT VALUES FOR THE
SMOOTHING CNSTANT (ALPHA).

AFTER EACH CHOICE OF ALPHA THE STANDARD ERROR OF THE
RESIDUALS WILL BE GIVEN.

IT IS ASSUMED THAT THERE IS VERY LITLE TREND PRESENT
IN THE DATA. THE INITIAL FORECAST EQUAL TO THE FIRST DATA
POINT WILL BE GENERATED BY THE PROGRAM.

HOW MANY DATA POINTS DO YOU HAVE? N= ? 16
ENTER THE ACTUAL VALUES ONE AT A TIME FOLLOWED BY A RETURN.
?

RESIDUALS WILL BE GIVEN.

IT IS ASSUMED THAT THERE IS VERY LITLE TREND PRESENT
IN THE DATA. THE INITIAL FORECAST EQUAL TO THE FIRST DATA
POINT WILL BE GENERATED BY THE PROGRAM.

HOW MANY DATA POINTS DO YOU HAVE? N= ? 16
ENTER THE ACTUAL VALUES ONE AT A TIME FOLLOWED BY A RETURN.
? 19
? 15
? 39
? 102
? 90
? 29
? 90
? 46
? 30
? 66
? 80
? 89
? 82
? 17
? 26
? 29

```
THE MEAN OF THE DATA POINTS IS 53.0625
WHAT ALPHA DO YOU WISH TO TRY?   ALPHA =   ? .1



THE STANDARD ERROR OF THE DEVIATIONS IS  33.58698

THE MSE IS  1332.518

DO YOU WISH TO TRY A DIFFERENT ALPHA?   TYPE 1 FOR YES
AND 0 (ZERO) FOR NO.      RESPONSE? 1




THE MEAN OF THE DATA POINTS IS 53.0625
WHAT ALPHA DO YOU WISH TO TRY?   ALPHA =   ? .3

THE STANDARD ERROR OF THE DEVIATIONS IS  34.45941

THE MSE IS  1136.064

DO YOU WISH TO TRY A DIFFERENT ALPHA?   TYPE 1 FOR YES
AND 0 (ZERO) FOR NO.      RESPONSE? 1




THE MEAN OF THE DATA POINTS IS 53.0625
WHAT ALPHA DO YOU WISH TO TRY?   ALPHA =   ? .5

THE STANDARD ERROR OF THE DEVIATIONS IS  34.5057

THE MSE IS  1119.354

DO YOU WISH TO TRY A DIFFERENT ALPHA?   TYPE 1 FOR YES
AND 0 (ZERO) FOR NO.      RESPONSE? 1




THE MEAN OF THE DATA POINTS IS 53.0625
WHAT ALPHA DO YOU WISH TO TRY?   ALPHA =   ? .7

THE STANDARD ERROR OF THE DEVIATIONS IS  34.99336

THE MSE IS  1148.81
```

```
DO YOU WISH TO TRY A DIFFERENT ALPHA?    TYPE 1 FOR YES
AND 0 (ZERO) FOR NO.        RESPONSE? 0



FORECAST( 1  ) = 19
FORECAST( 2  ) = 19
FORECAST( 3  ) = 16.2
FORECAST( 4  ) = 32.16
FORECAST( 5  ) = 81.048
FORECAST( 6  ) = 87.3144
FORECAST( 7  ) = 46.49432
FORECAST( 8  ) = 76.9483
FORECAST( 9  ) = 55.28449
FORECAST( 10 ) = 37.58535
FORECAST( 11 ) = 57.4756
FORECAST( 12 ) = 73.24268
FORECAST( 13 ) = 84.27281
FORECAST( 14 ) = 82.68184
FORECAST( 15 ) = 36.70455
FORECAST( 16 ) = 29.21137


F( 17  ) = 29.06341


DO YOU WISH TO TRY ANOTHER ALPHA?
AGAIN, TYPE 1 FOR YES AND 0 FOR NO.     RESPONSE? 0




THIS IS A SINGLE EXPONENTIAL SMOOTHING PROGRAM THAT WILL
ALLOW THE USER TO SIMULATE WITH DIFFERENT VALUES FOR THE
SMOOTHING CNSTANT (ALPHA).

AFTER EACH CHOICE OF ALPHA THE STANDARD ERROR OF THE
RESIDUALS WILL BE GIVEN.

IT IS ASSUMED THAT THERE IS VERY LITLE TREND PRESENT
IN THE DATA.   THE INITIAL FORECAST EQUAL TO THE FIRST DATA
POINT WILL BE GENERATED BY THE PROGRAM.



HOW MANY DATA POINTS DO YOU HAVE?   N= ? 20
ENTER THE ACTUAL VALUES ONE AT A TIME FOLLOWED BY A RETURN.
? 48
? 59
? 58
? 62
? 50
? 49
? 57
? 81
? 78
? 73
? 58
? 43
? 50
? 44
? 27
? 38
? 37
? 74
? 64
? 59
```

```
THE MEAN OF THE DATA POINTS IS 55.45
WHAT ALPHA DO YOU WISH TO TRY?  ALPHA =   ? 1

THE STANDARD ERROR OF THE DEVIATIONS IS  13.35931

THE MSE IS   169.85


DO YOU WISH TO TRY A DIFFERENT ALPHA?  TYPE 1 FOR YES
AND 0 (ZERO) FOR NO.       RESPONSE? 1




THE MEAN OF THE DATA POINTS IS 55.45
WHAT ALPHA DO YOU WISH TO TRY?  ALPHA =   ? .9

THE STANDARD ERROR OF THE DEVIATIONS IS  13.32661

THE MSE IS   169.1312


DO YOU WISH TO TRY A DIFFERENT ALPHA?  TYPE 1 FOR YES
AND 0 (ZERO) FOR NO.       RESPONSE? 1
```

```
THE MEAN OF THE DATA POINTS IS 55.45
WHAT ALPHA DO YOU WISH TO TRY?   ALPHA =  ? .8

THE STANDARD ERROR OF THE DEVIATIONS IS  13.3967

THE MSE IS  171.0701


DO YOU WISH TO TRY A DIFFERENT ALPHA?   TYPE 1 FOR YES
AND 0 (ZERO) FOR NO.       RESPONSE? 0

FORECAST(  4 ) = 57.76
FORECAST(  5 ) = 61.152
FORECAST(  6 ) = 52.2304
FORECAST(  7 ) = 49.64608
FORECAST(  8 ) = 55.52922
FORECAST(  9 ) = 75.90585
FORECAST( 10 ) = 77.58117
FORECAST( 11 ) = 73.91623
FORECAST( 12 ) = 61.18325
FORECAST( 13 ) = 46.63665
FORECAST( 14 ) = 49.32733
FORECAST( 15 ) = 45.06547
FORECAST( 16 ) = 30.6131
FORECAST( 17 ) = 36.52262
FORECAST( 18 ) = 36.90452
FORECAST( 19 ) = 66.5809
FORECAST( 20 ) = 64.51618


F( 21 ) = 60.10324


DO YOU WISH TO TRY ANOTHER ALPHA?
AGAIN, TYPE 1 FOR YES AND 0 FOR NO.      RESPONSE? 0
Ok
```

18

NONPARAMETRIC METHODS

CHAPTER LEARNING OBJECTIVES

On completion of this chapter the student should be able to

- Distinguish between parametric and nonparametric hypothesis tests.

- Calculate the sign test statistic, state the hypothesis, and carry out the sign test.

- Calculate the signed-rank statistic, state the hypothesis, and carry out the signed-rank test.

- Calculate the rank sum statistic, state the hypothesis, and carry out the rank sum test.

- Calculate the Spearman rank correlation, state the hypothesis, and carry out the rank correlation test.

SYNOPSIS

Most of the statistical tests covered so far have concerned population parameters. Population parameters are summary in nature and, as such, involve many assumptions when statistical tests concerning them are undertaken. Such assumptions as normality and homogeneity of variances are necessary to carry out parametric tests. When the assumptions are reasonable, parametric tests are the most powerful statistical tests available. However, situations arise where it is not feasible or economical to use parametric tests. Their use under these conditions results in invalid tests. In these cases, a class of statistical tests called nonparametric tests should be used instead. These tests generally involve a much simpler set of assumptions than parametric statistics and thus are applicable under a wider variety of conditions. This favorable aspect is offset by the fact that a larger sample size is needed to obtain the same level of confidence in the inferences. Additionally in many cases, the interpretation of the results is less useful than those made with parametric tests.

Nonparametric tests generally require no assumption about the form of the population probability distribution. For this reason, they are sometimes called distribution-free tests. This title may be misleading because even though no assumption is made about the form of the population, probability distributions still play a key role in hypothesis testing. In this case, they are introduced in such a way that the definition or procedures for calculating the statistics ensure that the probability distribution is appropriate.

Nonparametric tests involve hypotheses about the distribution of values or the comparison of the distributions of two populations. Conversely, parametric tests involve tests about the population parameters. Indirectly, population distributions were tested with parametric tests by the fact that if we concluded that two populations had equal means, other assumptions inferred that the populations also had the same distributions.

In the case of nonparametric tests, our statistics will concern the rank or order of the data. The statistics used to test the hypothesis generally assume, as an underlying idea, that order will occur randomly throughout a set of data. This proposition, actually the null hypothesis, is rejected when some pattern in the ranks or order is found. For example, when we find two variables in which the ranks appear to have similar values for the same experimental units, we reject the hypothesis of a random pattern and accept the alternative that a relationship exists between the two variables.

The Sign Test

The first technique described is the sign test. It involves the hypothesis that the distribution of data for two variables is the same. The procedure simply involves coding the data plus or minus, depending on whether variable A is larger than variable B. If the distributions are the same, the distribution of the plus and minus values should then follow a binomial probability distribution with $P = .5$. The sign test is a test of the hypothesis that $P = .5$.

The Signed-rank Test

The signed-rank test involves the same conditions as the sign test; only here the absolute magnitude of the differences is ranked. Signs are reattached to the rankings and either positive or negative ranks are summed. The expected value of this sum, if there is no difference in the population, is

$$E(T) = \frac{n(n+1)}{4}.$$

It can be tested with a Z value by

$$Z = \frac{T - E(T)}{\sqrt{\frac{n(n+1)(2n+1)}{24}}}$$

where T is the summed rank score.

The Rank Sum Test

The rank sum test is a further extension of the signed-rank test applied in situations where more than two populations are tested. It is somewhat analogous to analysis of variance. Data from all groups are collectively ranked. If there is no difference in the groups, the ranks then should be distributed randomly among all groups and the sum of ranks for each group should be equal. This hypothesis can be tested with the H statistic.

$$H = \frac{12}{n(n+1)} \sum_{K=1}^{K} \frac{R^2}{n_K} - 3(N+1)$$

when K = number of groups. The degrees of freedom are K - 1 and the distribution used is the chi-square. Values larger than the critical value lead to rejection.

The Rank Correlation Test

The rank correlation test is similar to the earlier use of the correlation coefficient in that, in general, both are measures of association. They differ, however, in the sense that rank correlation does not imply a linear relationship as does the correlation coefficient. Neither is the explanation of r^2 valid in rank correlation as the percentage of explained variance. Rank correlation provides an indication of a relationship that tends to persist throughout the data, as well as an indication of the direction of the relationship. A rank correlation and a correlation coefficient are not comparable values even though both vary between -1 and +1.

Rank correlation involves ranking the data for the first variable and then separately ranking the data for the second variable. Basically, although the appearance is different, an ordinary correlation coefficient is then computed between the rank scores of the two variables.

$$R = 1 - \frac{6(\sum_{i=1}^{n} D_i^2)}{n(n^2 - 1)}$$

The test statistic is a t value with n - 2 degrees of freedom and is calculated by

$$T = \frac{R\sqrt{n-2}}{\sqrt{1-R^2}}.$$

KEY TERMS

Distribution free   Characteristic of nonparametric methods that stem from the fact that they require no assumption about the pattern of the underlying population distribution.

Kruskal-Wallis test — A rank sum test that is suitable for testing whether various population distributions are identical. The only assumption required is that the populations are continuous.

Nonparametric methods — Methods used when measurements or observations are made in terms of ranks or orders.

Order techniques — Another name for nonparametric methods.

Parametric techniques — Hypothesis testing concerned with population parameters.

Sign test — Used when testing the hypothesis that two population distributions are identical. Sign refers to the negative in the number of values in one sample that are less than the number in the other. The test ignores the magnitude of differences.

Signed-rank test — Rank comparison test that considers the magnitude of differences.

Spearman's rank — Nonparametric method for testing the correlation between two variables. Requires no assumptions about the distributions of the underlying population.

FORMULA REVIEW

(18-1) Expected value of the random variable T

$$E(T) = \frac{n(n+1)}{4} \text{ when } n \geq 8$$

(18-2) Standard deviation of T

$$\sigma_T = \sqrt{\frac{n(n+1)(2n+1)}{24}}$$

(18-3) Z score for T

$$Z = \frac{T - E(T)}{\sigma_T}$$

(18-4) Test statistic H

$$H = \frac{12}{N(N+1)} \sum_{k=1}^{K} \frac{R_K^2}{n_K} - 3(N+1)$$

Distributed normally as X^2 with $(K-1)$ degrees of freedom.

(18-5) Spearman rank correlation coefficient

$$R = 1 - \frac{6 \sum_{i=1}^{n} D_i^2}{n(n^2 - 1)}$$

(18-6)   Spearman test
          statistic T $T = \dfrac{R\sqrt{n-2}}{\sqrt{1-R^2}}$

          Degrees of freedom (n - 2)

ORGANIZED LEARNING QUIZ

1. In the event of tied scores when calculating Spearman R, the average of the successive ranks that would have been assigned had the values been different is used.

   a. True
   b. False

2. Nonparametric methods can be used when

   a. observations are made in terms of ranks or orders.
   b. the nature of the distributions of underlying populations is unknown.
   c. nothing is known about population parameters.
   d. all the above.

3. A nonparametric test yields results as precise and consequently is as powerful a tool as a parametric technique.

   a. True
   b. False

4. For nonparametric methods, hypotheses may not be stated in the form $H_0: \mu_1 = \mu_2$

   a. True
   b. False

5. When nonparametric methods are used, we test the hypothesi that

   a. the population parameters differ.
   b. the sample proportions differ.
   c. the population distributions are identical.

6. If the null hypothesis that two distributions are the same is true, the differences between paired values will have

   a. mostly positive signs.
   b. mostly negative signs.
   c. the same number of positive as negative signs.
   d. there is insufficient information to complete the statement.

7. When the difference between two pairs is zero, that pair is disregarded and n is reduced.

   a. True
   b. False

8. The Kruskal-Wallis test requires that an assumption of non-normality of the population distributions be made.

    a. True
    b. False

9. The rank sum test requires _____ observations for each sample.

    a. five
    b. six
    c. eight

10. When using the Spearman rank correlation method,

    a. no assumption of normality is made.
    b. the relationship between variables is based on squared differences.
    c. the relationship between variables is determined on the basis of rank.
    d. a and c

11. The principal weakness in the sign test lies in the fact that _____.

12. The signed-rank test requires the ranking of all _____ between paired values from the _____ to the _____.

13. When the number of pairs considered in a signed-rank test is _____ or larger, the distribution of variable T is approximately normal.

14. The nonparametric method especially suited to testing whether population distributions are identical is the _____ _____ test.

15. Where possible, _____ techniques are to be preferred over _____ methods.

16. Describe the conditions under which nonparametric methods are used.

    _____

    _____

    _____

17. Explain why the signed-rank test is generally a more powerful test than the sign test.

    _____

    _____

    _____

18. Explain why the signed-rank test is a one-tailed test.

19. Discuss the advantages of using the Spearman rank correlation method.

20. Explain the use of the Spearman rank correlation method.

ANSWERS TO ORGANIZED LEARNING QUIZ

1. a
2. d
3. b
4. a
5. c
6. c
7. a
8. b   no assumptions about the distribution need be made.
9. a
10. d

11. It ignores the magnitude of the differences.

12. absolute difference, smallest, largest

13. eight

14. Kruskal-Wallis, or rank sum

15. parametric, nonparametric

16. Nonparametric methods can be used where no assumptions about the shape of the population distribution are made; they should be used when population parameters are unknown or when observations are in the form of rankings.

17. Under most circumstances the absolute magnitude of difference is important.

18. The H statistic is a function of the differences between distributions. When H is small, differences are small and the null hypothesis is not rejected. When H is sufficiently large, populations are not identical and the null hypothesis can be rejected.

19. No assumptions about the distributions of the underlying population are required; it avoids the elaborate computation

needed for the correlation coefficient r.

20. a. The scores for each of two variables are ranked.
    b. One of the two is arranged in ascending or descending order.
    c. The differences are calculated; differences are also squared.
    d. The sum of the squared differences is computed.
    e. R is calculated.
    f. T is calculated.

SAMPLE EXERCISE

Two managers have been asked to rank foremen on the same criteria. The results of their rankings are shown below.

Foreman	Manager 1	Manager 2
Billie	6	4
Randy	4	1
Gary	3	6
Dean	1	7
Judy	2	5
Al	7	8
Jerry	9	10
Ken	8	9
George	10	3
Sheel	5	2

Do the managers seem to agree?

The first step is to select a method. We will try the Spearman rank correlation. When using this method, we would normally rearrange one column in ascending or descending order.

Foreman	Manager 1	Manager 2	D	D^2
Dean	1	7	-6	36
Judy	2	5	-3	9
Gary	3	6	-3	9
Randy	4	1	3	9
Sheel	5	2	3	9
Billie	6	4	2	4
Al	7	8	-1	1
Ken	8	9	-1	1
Jerry	9	10	-1	1
George	10	3	7	49
			0	128

Differences are then calculated and squared, as in the column above. R is next calculated.

$$R = 1 - \frac{6 \Sigma D^2}{n(n^2 - 1)} = 1 - \frac{6(128)}{10(99)} = 1 - \frac{768}{990} = 1 - .7758 = .2242$$

$$T = \frac{R\sqrt{n-2}}{\sqrt{1-R^2}} = \frac{.2242\sqrt{10-2}}{\sqrt{1-(.2242)^2}} = \frac{.634}{.995} = .6503$$

The null hypothesis to be tested here is that there is no correlation between the two rankings; the alternate hypothesis is that there is. It is a two-tailed test, since we reject the null hypothesis if R is sufficiently close to either +1 or -1. The degrees of freedom are 10 - 2 = 8; if $\alpha = .01$, the critical t values from Appendix E are 3.355. The null hypothesis is not rejected, for there does not seem to be a significant relationship between the two rankings.

CASE PROBLEM

Portfolio Analysis

Security analysts frequently compare different securities for possible inclusion in investor's portfolios.

Portfolios are sets of securities and other assests into which investors place their funds.

Listed below are the performances of two different securities as well as the average market return.

Year	Returns Security A Blue Sky, Inc.	Security B Movers and Shakers, Inc.	(M) Market Returns
1965	.10	.07	.06
1966	.08	.09	.05
1967	.17	.11	.08
1968	-.11	.03	-.07
1969	.01	.01	-.02
1970	.17	.11	.14
1971	.15	.11	.11
1972	-.09	.06	.04
1973	.10	.08	.09
1974	.06	.09	.06
1975	-.21	.05	-.06
1976	.03	-.09	-.01
1977	.16	.10	.08
1978	.09	.11	.13

1. Compare the two securities. Is there any difference?

2. If so, compare the better security with the market returns. Did the security perform better than the market?

COMPUTER PROGRAMS

The BASIC program given covers all tests in this chapter.

For the sign test and the signed-rank test, the program requires as input the sample size and a pair of observations (X and y values) for each sample unit. The program computes the sign of the difference and counts the number of negative signs. The binomial probability of this result if p = .5 is then calculated. This is the sign test. For the signed-rank test, the program computes the absolute value of differences between the two variables. These data are ranked and the sign of the difference is attached to the rank. All negative ranks are summed and the appropriate Z statistic calculated. The program is dimensioned for 100 observations. A change of the dimension statement is all that is required to change dimensions.

For the Kruskall-Wallis test for the rank sum test, the program requires as input the number of groups, the sample size of each group, and the observations in each group. The program ranks the data collectively and then computes the sum-of-rank scores for each group. The H statistic, as defined in the text, is then computed. The program is dimensioned for 500 total observations and ten groups. Change only the dimension statement if this is to be changed.

For the Spearman rank correlation, the program requires as input the sample size and a pair of observations (X and y values) for each sample unit. The program ranks the data for each variable separately and then computes a correlation between the rank scores of the two variables. The program is deminsioned for 50 observations and can be changed by simply changing the dimension statement.

```
10 REM************************************************************
20 REM
30 REM        TITLE:     NONPARAM
40 REM        LANGAUGE:  BASIC
50 REM        DESCRIPTION:
60 REM           THIS PROGRAM CONTAINS THREE NONPARAMETRIC TESTS
70 REM           1) PROGRAM THAT COMPUTES THE SIGN TEST AND THE
80 REM              SIGNED-RANK TEST.
90 REM           2) PROGRAM THAT COMPUTES THE KRUSKALL-WALLIS
100 REM             TEST FOR THE RANK SUM TEST.
110 REM          3)  PROGRAM THAT COMPUTES THE SPEARMAN RANK
120 REM             CORRELATION.
130 REM       INSTRUCTIONS:
140 REM          AFTER ENTERING THE DESIRED PROGRAM, ENTER DATA
150 REM          AS REQUESTED.
160 REM
170 REM
180 REM************************************************************
190 REM
200 KEY OFF
210 DIM X(500),Y(100),R(500),V(100),S(100),B(10),W(2,50),Q(2,50)
220 CLS:LOCATE 8,1
230 PRINT "          THIS PROGRAM CONTAINS ALL OF THE NONPARAMETRIC"
240 PRINT "          TESTS COVERED IN CHAPTER 18 OF THE STUDENT"
250 PRINT "          GUIDE FOR STATISTICS FOR MANAGEMENT."
260 PRINT:PRINT
270 PRINT "                         MENU"
280 PRINT
290 PRINT "          1)   THE SIGN TEST AND THE SIGNED-RANK TEST"
300 PRINT
310 PRINT "          2)   THE KRUSKALL-WALLIS TEST FOR THE RANK"
320 PRINT "               SUM TEST"
330 PRINT
340 PRINT "          3)   THE SPEARMAN RANK CORRELATION COMPUTATION"
350 PRINT:PRINT
360 PRINT "WHICH PROGRAM DO YOU WISH?   (TYPE 1,2,OR 3 AND RETURN"
370 PRINT "                              OR TYPE 0 TO EXIT PROGRAM)"
380 INPUT P
390 CLS:PRINT
400 IF P=0 THEN 2380
410 IF P=1 THEN 450
420 IF P=2 THEN 1200
430 IF P=3 THEN 1780
440 IF P<>0 THEN 360
450 PRINT "PROGRAM TO COMPUTE THE SIGN TEST AND SIGNED-RANK TEST"
460 PRINT
470 PRINT
480 PRINT "PROBLEM NUMBER";
490 INPUT N
500 PRINT
510 PRINT "SAMPLE SIZE";
520 INPUT N
530 PRINT " INPUT OBSERVATIONS IN PAIRS SEPARATED BY COMMAS"
540 N1=0
550 FOR I=1 TO N
560 PRINT "OBS";I;
570 INPUT X(I),Y(I)
580 V(I)=Y(I)-X(I)
590 S(I)=0
600 IF V(I)=0 THEN 670
610 N1=N1+1
620 IF V(I)<0 THEN 650
630 S(I)=1
640 GOTO 660
650 S(I)=-1
660 V(I)=V(I)*S(I)
670 NEXT I
680 REM RANK V( ) VALUES
690 FOR I=1 TO 500
700 R(I)=0
710 NEXT I
720 FOR I=1 TO N
730 T1=0
740 T2=0
750 X1=V(I)
760 FOR J=1 TO N
770 IF X1<V(J) THEN 820
780 IF X1=V(J) THEN 810
790 T1=T1+1
800 GOTO 820
810 T2=T2+1
820 NEXT J
830 R(I)=T1+(T2+1)/2
```

```
840 NEXT I
850 T1=0
860 T2=0
870 FOR I=1 TO N
880 IF S(I)>=0 THEN 910
890 T1=T1+1
900 T2=T2+R(I)
910 NEXT I
920 PRINT
930 PRINT
940 PRINT "SIGN TEST"
950 PRINT "NEGATIVE SIGNS =";T1
960 GOSUB 1070
970 PRINT "P(X >= ";T1;CHR$(179);"P=.5, N=";N1;") =";P2
980 PRINT
990 PRINT "SIGNED-RANK TEST"
1000 E=N1*(N1+1)/4
1010 S4=(((N1*(N1+1))*(2*N1+1)/24))^.5
1020 PRINT "E(T)=";E;"    STD DEV=";S4;"    T=";T2
1030 Z=(T2-E)/S4
1040 PRINT:PRINT "Z= ";Z
1050 LOCATE 24,1:PRINT "TO CONTINUE PRESS ANY KEY":A$=INPUT$(1)
1060 GOTO 220
1070 P2=0
1080 FOR I=1 TO T1
1090 I1=I-1
1100 P=1
1110 IF I1=0 THEN 1150
1120 FOR J=1 TO I1
1130 P=P*(N1-J+1)/J
1140 NEXT J
1150 P=P*(.5^I1)*(.5^(N1-I1))
1160 P2=P2+P
1170 NEXT I
1180 P2=1-P2
1190 RETURN
1200 PRINT "PROGRAM TO COMPUTE KRUSKALL-WALLIS TEST"
1210 PRINT
1220 PRINT "PROBLEM NUMBER";
1230 INPUT N1
1240 PRINT
1250 PRINT
1260 FOR I=1 TO 10
1270 B(I)=0
1280 NEXT I
1290 PRINT "NUMBER OF GROUPS";
1300 INPUT N1
1310 N2=0
1320 FOR K=1 TO N1
1330 PRINT "SAMPLE SIZE OF GROUP";K;
1340 INPUT N(K)
1350 N2=N2+N(K)
1360 PRINT "OBSERVATIONS IN GROUP";K
1370 FOR I=N2-N(K)+1 TO N2
1380 PRINT "OBS";I-N2+N(K);
1390 INPUT X(I)
1400 NEXT I
1410 NEXT K
1420 REM RANK DATA
1430 FOR I=1 TO 500
1440 R(I)=0
1450 NEXT I
1460 FOR I=1 TO N2
1470 T1=0
1480 T2=0
1490 X1=X(I)
1500 FOR J=1 TO N2
1510 IF X1<X(J) THEN 1560
1520 IF X1=X(J) THEN 1550
1530 T1=T1+1
1540 GOTO 1560
1550 T2=T2+1
1560 NEXT J
1570 R(I)=T1+(T2+1)/2
1580 NEXT I
1590 REM SUM RANK SCORES
1600 FOR I=1 TO 10
1610 S(I)=0
1620 NEXT I
1630 N2=0
1640 S1=0
1650 FOR K=1 TO N1
```

```
1660 N2=N2+N(K)
1670 FOR I=N2-N(K)+1 TO N2
1680 S(K)=S(K)+R(I)
1690 NEXT I
1700 PRINT "SUM OF R FOR GROUP";K;"=";S(K)
1710 S1=S1+S(K)*S(K)/N(K)
1720 NEXT K
1730 PRINT
1740 H=(12/(N2*(N2+1)))*S1-3*(N2+1)
1750 PRINT "H=";H;"     DF=";N1-1
1760 LOCATE 24,1:PRINT "TO CONTINUE PRESS ANY KEY";A$=INPUT$(1)
1770 GOTO 220
1780 PRINT
1790 PRINT "PROGRAM TO COMPUTE SPEARMAN RANK CORRELATION"
1800 PRINT
1810 PRINT
1820 PRINT
1830 PRINT "PROBLEM NUMBER";
1840 INPUT N
1850 PRINT
1860 N1=2
1870 PRINT "SAMPLE SIZE";
1880 INPUT N
1890 PRINT "INPUT OBSERVATIONS IN PAIRS SEPARATED BY A COMMA"
1900 FOR I=1 TO N
1910 PRINT "OBS";I;
1920 INPUT W(1,I),W(2,I)
1930 NEXT I
1940 PRINT
1950 GOSUB 2160
1960 PRINT "OBS","RANK OF X","RANK OF Y","D SQUARED"
1970 D1=0
1980 D2=0
1990 FOR I=1 TO N
2000 D1=Q(1,I)-Q(2,I)
2010 D1=D1*D1
2020 D2=D2+D1
2030 PRINT I,Q(1,I),Q(2,I),D1
2040 NEXT I
2050 PRINT "    "," "," "," ","-------------"
2060 PRINT "    "," "," "," ",D2
2070 PRINT
2080 R1=1-6*D2/(N*(N*N-1))
2090 IF R1<>1 THEN 2120
2100 T1=1E+10
2110 GOTO 2130
2120 T1=R1*((N-2)^.5)/((1-R1*R1)^.5)
2130 PRINT "R=";R1;"   T=";T1;"   DF=";N-2
2140 LOCATE 24,1:PRINT "TO CONTINUE PRESS ANY KEY";:A$=INPUT$(1)
2150 GOTO 220
2160 REM -SUBROUTINE TO RANK DATA
2170 FOR I=1 TO 2
2180 FOR J=1 TO 50
2190 Q(I,J)=0
2200 NEXT J
2210 NEXT I
2220 FOR K=1 TO N1
2230 FOR I=1 TO N
2240 T1=0
2250 T2=0
2260 X1=W(K,I)
2270 FOR J=1 TO N
2280 IF X1<W(K,J) THEN 2330
2290 IF X1=W(K,J) THEN 2320
2300 T1=T1+1
2310 GOTO 2330
2320 T2=T2+1
2330 NEXT J
2340 Q(K,I)=T1+(T2+1)/2
2350 NEXT I
2360 NEXT K
2370 RETURN
2380 PRINT:PRINT "PROGRAM TERMINATED--THANK YOU"
2390 PRINT:PRINT:KEY ON:END
```

THIS PROGRAM CONTAINS ALL OF THE NONPARAMETRIC
TESTS COVERED IN CHAPTER 18 OF THE STUDENT
GUIDE FOR STATISTICS FOR MANAGEMENT.

```
                    MENU

        1)  THE SIGN TEST AND THE SIGNED-RANK TEST

        2)  THE KRUSKALL-WALLIS TEST FOR THE RANK
            SUM TEST

        3)  THE SPEARMAN RANK CORRELATION COMPUTATION
```

WHICH PROGRAM DO YOU WISH?   (TYPE 1,2,OR 3 AND RETURN
                              OR TYPE 0 TO EXIT PROGRAM)
? 1

PROGRAM TO COMPUTE THE SIGN TEST AND SIGNED-RANK TEST

PROBLEM NUMBER? 1


```
SAMPLE SIZE? 10
  INPUT OBSERVATIONS IN PAIRS SEPARATED BY COMMAS
OBS 1  ? 17,14
OBS 2  ? 17,21
OBS 3  ? 21,36
OBS 4  ? 18,20
OBS 5  ? 17,24
OBS 6  ? 14,12
OBS 7  ? 19,28
OBS 8  ? 17,16
OBS 9  ? 16,21
OBS 10 ? 12,20
```

SIGN TEST
NEGATIVE SIGNS = 3
P(X >= 3 P=.5, N= 10) = .9453125

SIGNED-RANK TEST
E(T)= 27.5   STD DEV= 9.810708   T= 7.5

Z= -2.038589
TO CONTINUE PRESS ANY KEY

THIS PROGRAM CONTAINS ALL OF THE NONPARAMETRIC
TESTS COVERED IN CHAPTER 18 OF THE STUDENT
GUIDE FOR STATISTICS FOR MANAGEMENT.


```
                    MENU

        1)  THE SIGN TEST AND THE SIGNED-RANK TEST

        2)  THE KRUSKALL-WALLIS TEST FOR THE RANK
            SUM TEST

        3)  THE SPEARMAN RANK CORRELATION COMPUTATION
```

WHICH PROGRAM DO YOU WISH?   (TYPE 1,2,OR 3 AND RETURN
                              OR TYPE 0 TO EXIT PROGRAM)
? 2

PROGRAM TO COMPUTE KRUSKALL-WALLIS TEST

PROBLEM NUMBER? 2

NUMBER OF GROUPS? 3

```
SAMPLE SIZE OF GROUP 1 ? 9
OBSERVATIONS IN GROUP 1
 OBS  1 ? 93
 OBS  2 ? 77
 OBS  3 ? 93
 OBS  4 ? 79
 OBS  5 ? 92
 OBS  6 ? 99
 OBS  7 ? 98
 OBS  8 ? 71
 OBS  9 ? 87
SAMPLE SIZE OF GROUP 2 ? 10
OBSERVATIONS IN GROUP 2
 OBS  1 ? 89
 OBS  2 ? 90
 OBS  3 ? 85
 OBS  4 ? 76
 OBS  5 ? 84
 OBS  6 ? 95
 OBS  7 ? 82
 OBS  8 ? 72
 OBS  9 ? 73
 OBS 10 ? 68


SAMPLE SIZE OF GROUP 3 ? 10
OBSERVATIONS IN GROUP 3
 OBS  1 ? 78
 OBS  2 ? 80
 OBS  3 ? 75
 OBS  4 ? 81
 OBS  5 ? 91
 OBS  6 ? 88
 OBS  7 ? 86
 OBS  8 ? 94
 OBS  9 ? 69
 OBS 10 ? 100
SUM OF R FOR GROUP 1 = 162
SUM OF R FOR GROUP 2 = 124
SUM OF R FOR GROUP 3 = 149

H= 2.051033      DF= 2
TO CONTINUE PRESS ANY KEY
```

THIS PROGRAM CONTAINS ALL OF THE NONPARAMETRIC
TESTS COVERED IN CHAPTER 18 OF THE STUDENT
GUIDE FOR STATISTICS FOR MANAGEMENT.

                    MENU

        1) THE SIGN TEST AND THE SIGNED-RANK TEST

        2) THE KRUSKALL-WALLIS TEST FOR THE RANK
            SUM TEST

        3) THE SPEARMAN RANK CORRELATION COMPUTATION

WHICH PROGRAM DO YOU WISH?   (TYPE 1,2,OR 3 AND RETURN
                              OR TYPE 0 TO EXIT PROGRAM)
 ? 3

PROGRAM TO COMPUTE SPEARMAN RANK CORRELATION

PROBLEM NUMBER? 3

SAMPLE SIZE? 12
INPUT OBSERVATIONS IN PAIRS SEPARATED BY A COMMA
OBS 1 ? 61.63
OBS 2 ? 64.62
OBS 3 ? 68.69
OBS 4 ? 69.65
OBS 5 ? 76.78
OBS 6 ? 78.70
OBS 7 ? 82.75
OBS 8 ? 84.90
OBS 9 ? 86.85
OBS 10 ? 90.84
OBS 11 ? 97.95
OBS 12 ? 98.86

OBS	RANK OF X	RANK OF Y	D SQUARED
1	1	2	1
2	2	1	1
3	3	4	1
4	4	3	1
5	5	7	4
6	6	5	1
7	7	6	1
8	8	11	9
9	9	9	0
10	10	8	4
11	11	12	1
12	12	10	4

			28

R= .9020979   T= 6.610605   DF= 10
TO CONTINUE PRESS ANY KEY

THIS PROGRAM CONTAINS ALL OF THE NONPARAMETRIC
TESTS COVERED IN CHAPTER 18 OF THE STUDENT
GUIDE FOR STATISTICS FOR MANAGEMENT.

                    MENU

       1)   THE SIGN TEST AND THE SIGNED-RANK TEST

       2)   THE KRUSKALL-WALLIS TEST FOR THE RANK
            SUM TEST

       3)   THE SPEARMAN RANK CORRELATION COMPUTATION

WHICH PROGRAM DO YOU WISH?   (TYPE 1,2,OR 3 AND RETURN
                              OR TYPE 0 TO EXIT PROGRAM)
? 0
OK

19

STATISTICAL QUALITY CONTROL

CHAPTER LEARNING OBJECTIVES

On completion of this chapter the student should be able to

- Distinguish between process control and acceptance sampling.
- Calculate a control chart for means ($\bar{X}$ chart).
- Calculate a control chart for ranges (R chart).
- Calculate a control chart for proportions defective (P chart).
- Calculate a control chart for number defectives (C chart).
- Define attribute sampling.
- Relate acceptance sampling to classical hypothesis testing.
- Define the terms producer's risk and consumer's risk.

SYNOPSIS

Statistical quality control is a good example of the application of statistics in a real-world situation. It is used extensively throughout business and government to answer questions on the quality of output of both goods and services. Its use has proved both economical and effective. The use of sampling to draw conclusions about overall quality is economical when large numbers of units are involved, a practical necessity if destructive tests are needed to measure quality, and more accurate in cases where inspection is tedious and prone to error.

Process Control for Variables

Statistical quality control can be divided into two general areas with distinct purposes. The area of process control is developed to draw conclusions as to whether a process is functioning properly. It is the attempt to draw a conclusion that concerns the future quality of the process. It does not attempt to draw conclusions concerning the units already produced. Acceptance sampling, on the other hand, is oriented toward historical quality. It attempts to answer the question as to whether a given batch of product meets a specific quality standard or not. Although differing in the time perspective of their conclusions, both process control and acceptance sampling use hypothesis testing for their underlying methodology.

Process control is developed on the basis that machines or production processes contain a certain amount of inherent or uncontrollable variation. It is quite feasible to measure this variation when the process is under control. If a measure reflects only uncontrollable variation, and later some additional variation is introduced, we should be able to detect this added variation through statistical means. Implied is the idea that the additional variation comes from controllable causes and, when detected, can be removed so that the process will once again be functioning, with the variation coming again only from uncontrollable factors. This same idea can be applied to several types of statistics and results in a variety of control charts. The most common control charts are the control chart for the mean, control chart for the range, the P chart, and the C chart.

The control charts can be regarded as repeated tests of hypotheses. The process is measured while producing good-quality products. These measurements are used to establish an upper control limit and a lower control limit. Such limits serve the same purpose as the critical value in hypothesis testing. Later sample results that fall outside the limits serve to reject the hypothesis that the process is producing in the same manner as when it was producing good quality. In this situation, the reason for this change in the process must be found and corrective action taken. The control limits are traditionally chosen at ± 3 standard deviations.

Process Control for Attributes/The C Chart

Each chart has a unique function and tends to detect different things. An $\bar{X}$ chart is most efficient in detecting a shift in the mean. An R chart will detect a change in the shape of the distribution. Together they can detect any shift in the probability distribution of the process.

The $\bar{X}$ and R charts are concerned with continuous measurements or with what is termed variable sampling. The P and C charts are concerned only with classification of something as good or bad (P chart) or with counting the number of defects (C chart). This process is termed attribute sampling. The P chart is a control chart on the proportion of defects of a sample. The C chart concerns counting the number of defectives for each sampling unit.

The basic process in generating a control chart is the same regardless of the type of chart. The only difference is the sample statistic involved. The $\bar{X}$ chart will be used to illustrate this process. Several samples are taken from a process during the time it is considered to be under control. The sample statistic,

$\bar{\bar{X}}$ (the mean of several sample means), is calculated along with the associated standard error. For an R chart, this is $\bar{R}$ and σ_R. For a C chart, it is $\bar{C}$ and σ_C. The upper control limit becomes the sample statistic, $\bar{\bar{X}}$ for an $\bar{X}$ chart, plus three times the standard deviation; and the lower control limit is the sample statistic, minus three times the standard deviation.

$$UCL_{\bar{X}} = \bar{\bar{X}} + 3\bar{\sigma}$$

$$LCL_{\bar{X}} = \bar{\bar{X}} - 3\sigma_{\bar{X}}$$

Once limits are established, periodic samples are taken and the test statistic calculated. In this example it would be $\bar{X}$. If the sample statistic falls outside the limits, we conclude that the process is out of control and appropriate action is taken.

Acceptance Sampling

Although the methods of hypothesis testing are not always clearly evident in process control, acceptance sampling is very simply a direct application. For simplicity, this discussion is confined to the test statistic--proportion of defectives. The ideas are applicable to other test statistics. In hypothesis testing of proportions, we must first identify a hypothesized mean. This is the value that is considered acceptable or good quality. In quality control it is termed the acceptable quality level (AOQ). Also included is a specified value of α, the type I error (also termed producer's risk). As the percentage of defectives increases above the value of what is considered acceptable quality, a gray region is entered in which quality is not good; but it is not bad enough to invoke the expense and trouble rejecting a batch of goods. During transition of this gray area, sooner or later a point is reached that we are willing to call bad quality. In quality control this point is termed the Lot Tolerance Percent Defective (LTPD). It can, in some sense, be regarded as an alternative hypothesis. In conjuction with this alternative hypothesis, there is the risk of wrongful acceptance of a batch of goods when the alternative hypothesis is, in fact, true. By definition, it is a type II error and the probability of this occurrence is β (also termed consumer's risk). The definitions of AOQ, α, LTPD, and β determine a unique sampling plan (n_1C). There is one combination of sample size n and critical value C that will satisfy the stated hypothesis test (the calculation of n_1C is somewhat tedious; so tables have been developed for this purpose). The procedure used is to take a sample of size n and test the items, classifying them as either good or bad. If the sample proportion of defectives is greater than C, the hypothesis of good quality is rejected; the total batch of goods is also rejected because it does not meet acceptable quality standards.

KEY TERMS

Acceptance sampling   Random sampling from a lot or process to determine if the lot or batch is acceptable.

Attribute	Quality characteristic of a product that cannot be expressed as a continuous variable, such as "good" or "defective."
C Chart	Chart on which is plotted the number of defects per sampled unit for many units taken from a manufacturing process.
Centerline	Horizontal line on control chart showing the computed mean.
Consumer's risk	Purchaser's risk of accepting a lot that is, in fact, bad; it is denoted by β (beta) and is called a type II error.
Control band	Range of allowable values for the process in question. When measurements fall either above or below this range, the process is considered out of control.
Critical value	Predetermined number to which the number of defects is compared. If the number of rejects is less than the acceptance number or critical value, the lot is accepted.
Decision rule	Criterion used to determine or guide the decision to accept or reject a lot.
Lower control line	Line drawn parallel to and below the centerline on a control chart. The line forms the lower limit of the control band.
Mean chart	Type of control chart designed to show variations of the sample mean from the grand mean, the overall mean of the measurements taken.
Producer's risk	Risk that a buyer will reject a lot when it is good; it is denoted by α (alpha) and is called a type I error.
Proportion defective chart	Control chart on which the proportion of defects in each sample is plotted.
Range chart	Range refers to the difference between the smallest and largest measurements in a sample. The range chart is used to show the fluctuations of the ranges of the samples about the average ($\bar{R}$) of the ranges for all samples.
Sample number	Number assigned to sample, which is then entered on a control chart.
States of nature	The two possible conditions for a given lot or batch: good (acceptable) or defective.
Statistical sampling plan	Plan designating the number of units that must be selected from a lot of given lot

	size.
Time point	The horizontal axis on a control chart shows the point in time at which a sample is taken.
Upper control limit	Line drawn parallel to and above the centerline on a control chart. It forms the upper limit of the control band.
Variations from specifications	Variations are of two types: assignable, those that can be identified and thus controlled, and random.

FORMULA REVIEW

(19-1) $\bar{X}$ chart $\qquad UCL_{\bar{X}} = \bar{\bar{X}} + 3S_{\bar{X}} = \bar{\bar{X}} + 3(S/\sqrt{n})$

(19-2) $\qquad LCL_{\bar{X}} = \bar{\bar{X}} - 3S_{\bar{X}} = \bar{\bar{X}} - 3(S/\sqrt{n})$

(19-5) Standard deviation R chart $\qquad S_R = \sqrt{\dfrac{\Sigma(R-\bar{R})^2}{m-1}}$

(19-4) R chart $\qquad UCL_R = \bar{R} + 3S_R$

$\qquad LCL_R = \bar{R} - 3S_R$

(19-6) Standard deviation P chart $\qquad \sigma_P = \sqrt{\dfrac{\bar{P}(1-\bar{P})}{n}}$

(19-7) P chart $\qquad UCL_P = \bar{P} + 3\sigma_P$

$\qquad LCL_P = \bar{P} - 3\sigma_P$

(19-8) Mean for C chart $\qquad \bar{C} = \dfrac{C_1 + C_2 \cdots + C_m}{m}$

(19-9) C chart $\qquad UCL_C = \bar{C} + 3\sqrt{\bar{C}}$

$\qquad LCL_C = \bar{C} - 3\sqrt{\bar{C}}$

ORGANIZED LEARNING QUIZ

1. The two types of variables in a manufacturing process are distinguished by

    a. whether they can be identified.
    b. whether they can be controlled.
    c. whether they occur regularly or infrequently.

2. A manufacturing process is said to be in control when

   a. all random variation is eliminated.
   b. 90% of all random variation is eliminated.
   c. assignable variations are entirely eliminated.
   d. assignable variations are brought within acceptable limits.

3. Quality control techniques are diagnostic rather than remedial.

   a. True
   b. False

4. Quality control techniques show, in a probabilistic sense only, whether a manufacturing process is meeting specifications.

   a. True
   b. False

5. Statistical quality control techniques require the use of descriptive statistics only.

   a. True
   b. False

6. The upper and lower control limits

   a. are set above and below the highest and lowest observed values.
   b. are set plus or minus one standard deviation from the centerline.
   c. are set plus or minus three standard deviations from the centerline.

7. The centerline is computed on the basis of

   a. where the manufacturing process ideally operates.
   b. the overall computed mean.
   c. the distance between the upper and lower control limits.

8. The mean chart is designed to

   a. show the average performance of the process at any point in time.
   b. show the mean variation in performance from the desired value.
   c. show the variation of the sample means from the grand mean.

9. The range chart is designed to show

   a. the highest and lowest observations of a process.
   b. the highest and lowest acceptable observations of a process.
   c. the fluctuations of the ranges of the samples about the average for all samples.

10. A P chart is

    a. a control chart showing the proportion of defective units.
    b. used to plot the total number of defects based on 100% sampling.

c. best suited for use with small samples.

11. Control charts have been developed for the control of _____ _____ and _____.

12. Quality is a product _____; size is a product _____.

13. The upper control limit on a mean chart is set at _____ _____ standard deviations above the centerline because _____
_____.

14. The C chart is a chart on which is plotted the number of defects per _____.

15. The distribution of the defects per unit of product closely approximates the _____ distribution.

16. The procedure used for screening the quality of incoming parts or raw materials is a _____.

17. The probability of making a type I error is called the _____ because _____
_____.

18. The probability of making a type II error is called the _____ because _____
_____.

19. Measurements going outside the control limits on the R chart indicate that _____.

20. Measurements going outside the control limits on the mean chart indicate that _____.

21. Define and explain the objectives of statistical quality control. _____
_____
_____
_____

22. Explain the construction and use of the mean chart. _____
_____
_____
_____

23. Explain the construction and use of the R chart. _____
_____
_____

24. Explain the construction and use of the C chart.

25. Explain the procedure of screening parts through the use of a statistical sampling plan.

ANSWERS TO ORGANIZED LEARNING QUIZ

1. b
2. d
3. a
4. a
5. b
6. c
7. b
8. c
9. c
10. a and sometimes b

11. variables, attributes

12. attribute, variable

13. three; this will include almost all sample values.

14. unit of manufactured product.

15. Poisson

16. statistical sampling plan

17. producer's risk; it means a good lot has been rejected.

18. consumer's risk; it means a bad lot has been accepted.

19. Something is wrong with the process with respect to the variability.

20. Something is wrong with the process with respect to the mean.

21. Statistical quality control is an approach to manufacturing control that uses statistical techniques to measure variations and ranges in quality and to aid in decision making.

22. See section 19-1 of the textbook.

23. See section 19-1 of the textbook.

24. See section 19-3 of the textbook.

25. See section 19-4 of the textbook.

SAMPLE EXERCISE

The procedure in applying control charts may be divided into a number of distinct steps.

1. Selecting the process characteristics to be controlled.
2. Choosing the statistical measures to be used.
3. Collecting and organizing the data.
4. Establishing the control limits and making decision rules.
5. Reviewing the results and comparing with decision rules.

The Joy Juice Bottling Company wishes to apply the techniques of statistical quality control to their new bottling process. This exercise demonstrates the procedure in their plant, using the five steps outlined above.

1. The characteristic to be controlled is the volume of Joy Juice that goes into each 16-ounce bottle.

2. The statistical measures chosen might be the mean and range control charts.

3. Table: Joy Juice Bottling Process

Sample Number	Volume in Ounces of the Six Bottles in Each Carton						$\bar{X}$	ΣX^2	$\bar{X}^2$	S_x^2	R	$(R-\bar{R})^2$
	1	2	3	4	5	6						
1	15.9	15.8	16.0	15.8	16.3	16.1	16.0	1533.0	256.0	.5	.5	.16
2	15.6	15.7	15.3	15.0	16.1	16.1	15.6	1467.4	243.4	1.17	1.1	.04
3	16.1	15.3	15.4	15.7	16.2	16.1	15.8	1498.6	249.6	.167	.9	.00
4	16.4	16.0	15.7	16.1	15.0	15.9	15.9	1508.5	252.8	1.39	1.4	.25
5	16.1	15.8	16.3	16.0	15.4	16.0	15.9	1523.7	252.8	1.15	.9	.00
6	16.0	15.4	15.8	15.7	15.4	15.9	15.8	1479.3	249.6	3.05	.6	.09
Total							95.0	9010.5	1504.	7.427	5.4	.54

$\bar{R} = .9$

4. Calculations

a. Mean Chart

Grand Mean $\bar{\bar{X}} = \dfrac{\bar{X}_1 + \bar{X}_2 + \cdots + \bar{X}_6}{6} = \dfrac{95}{6} = 15.83$

Average population variance $s_X^2 = \dfrac{\Sigma s_X^2}{n} = \dfrac{7.427}{6} = 1.238$

Unbiased Estimator $s_X^2 = \dfrac{n}{n-1} \; s_X^2 = \dfrac{6}{6-1}(1.24) = 1.4856$

Variance of sample mean $s_{\bar{X}}^2 = \dfrac{1.4856}{6} = .2476$

Standard error = .4976

$UCL_{\bar{X}}$ = 16 + 3(.4976) = 17.4928

$LCL_{\bar{X}}$ = 16 − 3(.4976) = 14.5072

b. Range Chart

$$S_R = \sqrt{\frac{\Sigma (R - \bar{R})^2}{m - 1}} = \sqrt{\frac{.54}{5}} = .329$$

UCL_R = .9 + 3(.329) = 1.887

LCL_R = .9 − 3(.329) = −.087 = 0

5a. The mean chart could be drawn up and the sample means plotted. By looking at the data, it can be seen that none of the sample means falls outside the control band.

b. By examining the data, it can be seen that none of the ranges falls outside the control band.

CASE PROBLEM

Microsystems Division, Moon Industries

Quality control records maintained by the manufacturing department of Microsystems Division revealed the following data on subassembly 3 of the new product Micro-Mooner, based on the first shifts of the first three weeks of production.

Day	Defective Parts Replaced (%)	Number of Adjustments Required per Hour
1	2.3	20
2	1.8	36
3	1.9	18
4	2.3	30
5	2.5	24
6	2.1	20
7	1.7	27
8	1.6	22
9	2.0	40
10	1.8	33
11	2.3	18
12	2.1	27
13	2.2	26
14	1.9	20
15	1.9	32

Merlin Moon, manager of the Microsystems division, needs to know how the production process on his new product is going. As an aid in keeping replacement costs and mechanical adjustments to a minimum, calculate and prepare appropriate control charts based on three-sigma limits.

COMPUTER PROGRAM

Program SQC is provided in this chapter to compute the control chart for means and the control chart for ranges. The program requires as input the sample size, the number of observations, and the observation for each sample. The program computes the grand mean and its standard error and the average range and its standard error. Control charts are then computed and a plot of both charts is output. The program is dimensioned for A(I, J), where I is the maximum number of samples and J is the maximum number of observations plus 5. Only the dimension statement need be changed to change the dimensions for the program.

```
10 REM***********************************************************
20 REM
30 REM      TITLE:      SQC
40 REM      LANGAUGE:   BASIC
50 REM      DESCRIPTION:
60 REM         PROGRAM SQC COMPUTES THE CONTROL CHARTS FOR
70 REM         MEANS AND RANGES. REQUIRED INPUTS ARE THE
80 REM         SAMPLE SIZE, THE NO. OF OBSERVATIONS, AND
90 REM         THE OBS. FOR EACH SAMPLE.  THE PROGRAM IS
100 REM        DIMENSIONED FOR A(I,J) WHERE I IS THE MAXI-
110 REM        MUM NO. OF SAMPLES AND J IS THE MAXIMUM NO.
120 REM        OF OBS. PLUS 5.  ONLY THE DIMENSION STATE-
130 REM        MENT NEED BE CHANGED TO CHANGE THE DIMENSIONS
140 REM        FOR THE PROGRAM.
150 REM     INSTRUCTIONS:
160 REM        ENTER DATA AS REQUESTED BY THE PROGRAM.
170 REM
180 REM***********************************************************
190 REM
200 KEY OFF
210 DIM A(21,25),C$(75),B(75),B$(75)
220 FOR I=1 TO 75
230 B$(I)=CHR$(196)
240 NEXT I
250 FOR I=1 TO 21
260 FOR J=1 TO 25
270 A(I,J)=0
280 NEXT J:NEXT I
290 CLS
300 PRINT "                PROGRAM TO COMPUTE CONTROL CHARTS"
310 PRINT
320 PRINT
330 PRINT "PROBLEM NUMBER";
340 INPUT N
350 PRINT
360 PRINT
370 PRINT "SAMPLE SIZE";
380 INPUT N
390 PRINT "NUMBER OF SAMPLES";
400 INPUT N1
410 R1=0
420 X1=0
430 X2=0
440 R2=0
450 S1=0
460 X3=0
470 FOR I=1 TO N1
480 A(I,N+1)=0
490 A(I,N+2)=0
500 A(I,N+3)=1E+08
510 A(I,N+4)=0
520 PRINT "OBSERVATIONS FOR SAMPLE";I
530 FOR J=1 TO N
540 PRINT "OBS";J;
550 INPUT A(I,J)
560 A(I,N+1)=A(I,N+1)+A(I,J)
570 A(I,N+2)=A(I,N+2)+A(I,J)*A(I,J)
580 IF A(I,J)>A(I,N+3) THEN 600
590 A(I,N+3)=A(I,J)
600 IF A(I,J)<A(I,N+4) THEN 620
610 A(I,N+4)=A(I,J)
620 NEXT J
630 PRINT
640 A(I,N+4)=A(I,N+4)-A(I,N+3)
650 R1=A(I,N+4)+R1
660 R2=A(I,N+4)*A(I,N+4)+R2
670 A(I,N+1)=A(I,N+1)/N
680 A(I,N+3)=A(I,N+2)/N-A(I,N+1)*A(I,N+1)
690 S1=S1+A(I,N+3)
700 X1=X1+A(I,N+1)
710 X2=X2+A(I,N+2)
720 X3=X3+A(I,N+1)*A(I,N+1)
730 NEXT I
740 R5=R1/N1
750 REM--PRINT TABLE--
```

```
760  PRINT
770  PRINT
780  R4=0
790  PRINT "      -      "," "," "," "," "," - "
800  PRINT "      X       ","SUM OF X SQ","STD ERROR OF X";"   RANGE","(R-R)SQ"
810  FOR I=1 TO 75:PRINT B$(I);:NEXT I:PRINT
820  FOR I=1 TO N1
830  R3=A(I,N+4)-R5
840  R3=R3*R3
850  R4=R3+R4
860  PRINT A(I,N+1),A(I,N+2),A(I,N+3),A(I,N+4),R3
870  NEXT I
880  FOR I=1 TO 75:PRINT B$(I);:NEXT I:PRINT
890  PRINT X1,X2,S1,R1,R4
900  PRINT
910  LOCATE 24,1:PRINT "TO CONTINUE OUTPUT, PRESS ANY KEY";:A$=INPUT$(1)
920  PRINT
930  PRINT
940  REM--CALCULATE CHART--
950  S2=(((S1/N1)*(N/(N-1)))/N)^.5
960  C1=X1/N1+3*S2
970  C2=X1/N1-3*S2
980  S3=(R4/(N1-1))^.5
990  C3=R1/N1+3*S3
1000 C4=R1/N1-3*S3
1010 IF C4>=0 THEN 1030
1020 C4=0
1030 PRINT "UCLX=";X1/N1;"+3 X (";S2;")=";C1
1040 PRINT "LCLX=";X1/N1;"-3 X (";S2;")=";C2
1050 PRINT "UCLR=";R1/N1;"+3 X (";S3;")=";C3
1060 PRINT "LCLR=";R1/N1;"-3 X (";S3;")=";C4
1070    N5=N+1
1080    K=0
1090 GOSUB 1200
1100 N5=N+4
1110 K=1
1120 LOCATE 24,1:PRINT "TO CONTINUE OUTPUT, PRESS ANY KEY";:A$=INPUT$(1)
1130 PRINT
1140 GOSUB 1200
1150 LOCATE 24,1:PRINT "TO CONTINUE PRESS ANY KEY":A$=INPUT$(1)
1160 PRINT "ANOTHER PROBLEM ; 1=YES, 0=NO";:INPUT G
1170 IF G=1 THEN 220
1180 IF G=0 THEN 1730
1190 GOTO 1160
1200 REM --PLOT CONTROL CHARTS--
1210 REM --N5 IS INDEX TO VALUE PLOTTED--
1220 REM --K=0: CONTROL CHART FOR MEANS; K=1: CONTROL CHART FOR RANGES--
1230 PRINT:PRINT:PRINT
1240 IF K=1 THEN 1290
1250 IF K=1 THEN 1290
1260 M1=C2
1270 M2=C1
1280 GOTO 1310
1290 M1=C4
1300 M2=C3
1310 FOR I=1 TO N1
1320 IF A(I,N5)>M1 THEN 1340
1330 M1=A(I,N5)
1340 IF A(I,N5)<M2 THEN 1360
1350 M2=A(I,N5)
1360 NEXT I
1370 A1=((M2-M1)/60)*1.1
1380 IF K=1 THEN 1430
1390 A2=INT(((C1-M1)/A1)+.5)+3
1400 A3=INT(((C2-M1)/A1)+.5)+3
1410 A4=INT(((X1/N1-M1)/A1)+.5)+3
1420 GOTO 1460
1430 A2=INT(((C3-M1)/A1)+.5)+3
1440 A3=INT(((C4-M1)/A1)+.5)+3
1450 A4=INT(((R1/N1-M1)/A1)+.5)+3
1460 FOR I=1 TO N1
1470 FOR J=1 TO 75
1480 C$(J)=" "
1490 NEXT J
```

```
1500 I1=INT(((A(I,N5)-M1)/A1)+.5)+3
1510 C$(A4)="I"
1520 C$(A2)="I"
1530 C$(A3)="I"
1540 C$(I1)="*"
1550 FOR J=1 TO 75:PRINT C$(J);:NEXT J:PRINT
1560 NEXT I
1570 FOR I=1 TO 75
1580 PRINT B$(I);
1590 C$(I)=" "
1600 NEXT I
1610 PRINT
1620 C$(A3-1)="M-3S"
1630 C$(A2-4)="M+3S"
1640 C$(A4-3)="M"
1650 FOR I=1 TO 60:PRINT C$(I);:NEXT I:PRINT
1660 PRINT
1670 IF K=1 THEN 1700
1680 PRINT "CONTROL CHART FOR MEANS"
1690 GOTO 1720
1700 PRINT "CONTROL CHART FOR RANGES"
1710 PRINT:PRINT
1720 RETURN
1730 KEY ON: END
```

PROGRAM TO COMPUTE CONTROL CHARTS

PROBLEM NUMBER? 1

SAMPLE SIZE? 5
NUMBER OF SAMPLES? 8
OBSERVATIONS FOR SAMPLE 1
OBS 1 ? 42
OBS 2 ? 44
OBS 3 ? 45
OBS 4 ? 47
OBS 5 ? 42

OBSERVATIONS FOR SAMPLE 2
OBS 1 ? 44
OBS 2 ? 46
OBS 3 ? 48
OBS 4 ? 47
OBS 5 ? 45

OBSERVATIONS FOR SAMPLE 3
OBS 1 ? 40
OBS 2 ? 41
OBS 3 ? 42
OBS 4 ? 44
OBS 5 ? 48

OBSERVATIONS FOR SAMPLE 4
OBS 1 ? 47
OBS 2 ? 48
OBS 3 ? 46
OBS 4 ? 45
OBS 5 ? 44

OBSERVATIONS FOR SAMPLE 5
OBS 1 ? 47
OBS 2 ? 44
OBS 3 ? 46
OBS 4 ? 44
OBS 5 ? 39

```
OBSERVATIONS FOR SAMPLE 6
OBS 1 ? 38
OBS 2 ? 47
OBS 3 ? 45
OBS 4 ? 46
OBS 5 ? 44

OBSERVATIONS FOR SAMPLE 7
OBS 1 ? 50
OBS 2 ? 46
OBS 3 ? 47
OBS 4 ? 47
OBS 5 ? 50

OBSERVATIONS FOR SAMPLE 8
OBS 1 ? 42
OBS 2 ? 48
OBS 3 ? 46
OBS 4 ? 44
OBS 5 ? 45
```

$\bar{X}$	SUM OF X SQ	STD ERROR OF X	RANGE	$(R-\bar{R})$SQ
44	9698	3.599976	5	1
46	10590	2	4	4
43	9285	8	8	4
46	10590	2	4	4
44	9718	7.599976	8	4
44	9730	10	9	9
48	11534	2.800049	4	4
45	10145	4	6	0
360	81290	40	48	30

```
TO CONTINUE OUTPUT, PRESS ANY KEY


UCLX= 45 +3 X ( 1.118034 )= 48.3541
LCLX= 45 -3 X ( 1.118034 )= 41.6459
UCLR= 6 +3 X ( 2.070197 )= 12.21059
LCLR= 6 -3 X ( 2.070197 )= 0


        I               *   I                   I
        I                   I       *           I
        I       *           I                   I
        I                   I   *               I
        I           *       I                   I
        I           *       I                   *
        I                   I                   I
        I                   *                   I

      M-3S                  M                 M+3S

CONTROL CHART FOR MEANS
TO CONTINUE OUTPUT, PRESS ANY KEY


        I                *  I                   I
        I           *       I                   I
        I                   I   *               I
        I       *           I                   I
        I                   I       *           I
        I                   I           *       I
        I           *       I                   I
        I                   *                   I

      M-3S                  M                 M+3S

CONTROL CHART FOR RANGES


TO CONTINUE PRESS ANY KEY
ANOTHER PROBLEM ; 1=YES, 0=NO? 0
OK
```